# PROBLEMS IN ANALYTIC NUMBER THEORY

# Problems in Analytic Number Theory

*Editor*

Danyal Sadik

scitus academics

**Problems in Analytic Number Theory**
Edited by **Danyal Sadik**

ISBN: 978-1-68117-565-2
Library of Congress Control Number: 2016942995


SCITUS Academics LLC,
www.scitusacademics.com
Box No. 4766, 616 Corporate Way,
Suite 2, Valley Cottage,
NY 10989



**Notice**



Printed in United States of America on Acid Free Paper

# PREFACE

One might have thought that number theory was simply the study of numbers, but that is too broad a definition, since numbers are almost ubiquitous in mathematics. Number theory is a vast and fascinating field of mathematics, sometimes called "higher arithmetic," consisting of the study of the properties of whole numbers. Primes and prime factorization are especially important in number theory, as are a number of functions such as the divisor function, Riemann zeta function, and totient function. Analytic number theory is a branch of number theory that uses methods from mathematical analysis to solve problems about the integers. Analytic number theory, and its applications and interactions, are currently experiencing intensive progress, in sometimes unexpected directions. In recent years, many important classical questions have seen spectacular advances based on new techniques; conversely, methods developed in analytic number theory have led to the solution of striking problems in other fields. Recent advances in analytic number theory have had repercussions in various mathematical subjects, such as harmonic analysis, ergodic theory and dynamics, additive and multiplicative combinatorics and theoretical computer science.

The biggest technical change after 1950 has been the development of sieve methods, particularly in multiplicative problems. These are combinatorial in nature, and quite varied. The extremal branch of combinatorial theory has in return been greatly influenced by the value placed in analytic number theory on quantitative upper and lower bounds. Another recent development is probabilistic number theory, which uses methods from probability theory to estimate the distribution of number theoretic functions, such as how many prime divisors a number has.

*Problems in Analytic Number Theory* present a problem-solving approach to the difficult subject of analytic number theory. This book is focused at researchers, teachers, and graduate students interested in number theory and its links with other branches of science.

# TABLE OF CONTENTS

7

# UNICITY OF TYPES FOR SUPERCUSPIDAL REPRESENTATIONS OF *P*-ADIC $\mathbf{SL}_2$

**PeterLatham**

Department of Mathematics, The University of East Anglia, Norwich, UnitedKingdom

## ABSTRACT

We consider the question of unicity of types on maximal compact subgroups for supercuspidal representations of $\mathbf{SL}_2$ over a nonarchimedean local field of odd residual characteristic. We introduce the notion of an archetype as the $\mathbf{SL}_2$-conjugacy class of a typical representation of a maximal compact subgroup, and go on to show that any archetype in $\mathbf{SL}_2$ is restricted from one in $\mathbf{GL}_2$. From this it follows that any archetype must be induced from a Bushnell–Kutzko type. Given a supercuspidal representation $\pi$ of $\mathbf{SL}_2(F)$, we give an additional explicit description of the number of archetypes admitted by $\pi$ in terms of its ramification. We also describe a relationship between archetypes for $\mathbf{GL}_2$ and $\mathbf{SL}_2$ in terms of $L$-packets, and deduce an inertial Langlands correspondence for $\mathbf{SL}_2$.

## KEYWORDS

Bushnell–Kutzko types; ***p***-adic groups Special linear group; Langlands correspondence

## 1. INTRODUCTION

The local Langlands conjectures predict the existence of natural parametrizations of certain sets of representations of the Weil–Deligne group of a nonarchimedean local field $F$ and (packets of) smooth, irreducible representations of reductive groups defined over $F$. While this is of theoretical significance, in practice it can be rather difficult to obtain explicit information from this correspondence. The theory of types arose as a means of obtaining an explicit understanding of the representation theory of $p$-adic groups, partly in the hope that this would allow one to obtain explicit information on the Weil group side of the Langlands correspondence.

The idea of the theory of types is to find a family of irreducible representations $\lambda$ of certain compact open subgroups $J$ of a $p$ -adic group G such that the presence of $\lambda$ in the restriction to $J$ of an irreducible representation $\pi$ of G allows one to identify $\pi$ , at least as closely as such an approach may possibly do so. This then naturally ties in with ideas due to Bernstein, allowing one to interpret types as effective means of describing the blocks of the category Rep(G) of smooth representations of G. The existence of types then allows one, via the theory of Hecke algebras, to give an effective description of Rep(G).

The existence of types is now known in many situations. In each case, the approach is to first provide an explicit construction of those types which are contained in the supercuspidal representations of G. This has now been done when G is a general or special linear group [4], [5] and [6], when G is a classical group in odd residual characteristic [19], when G is an inner form of a general linear group [16] and [17], as well as for arbitrary G over fields of characteristic zero, modulo some assumptions on the base field which, in particular, are always satisfied for $p$ sufficiently large [20] and [11]. Via the Bushnell–Kutzko theory of covers [7] there is then a standard approach for extending these constructions to the non-cuspidal representations. Again, this has been completed in many cases: for general linear groups [8], special linear groups [10], classical groups in odd residual characteristic [13] and inner forms of general linear groups [18].

Call any type arising from such a construction a "Bushnell–Kutzko" type. It is simple to see that, when (J,λ) is a Bushnell–Kutzko type and $K$ is a compact subgroup of Gcontaining $J$, then the irreducible components of the representation obtained by inducing$\lambda$ to $K$ are also Bushnell–Kutzko types. To date, no other types are known to exist. It is therefore natural to conjecture that repeating this process for all Bushnell–Kutzko types(J,λ) and all such $K$ exhausts the set of isomorphism classes of types – this is what we refer to as the "unicity of types" property. This conjecture appears to be intimately related to several topics of interest in the Langlands program. In particular, it should imply (but be stronger than) the existence of a *unique* inertial Langlands correspondence, and in the case of $\mathbf{GL}_N$(F) has found multiple applications to problems in the spirit of the Breuil–Mézard conjecture.

It was in order to allow an application to the Breuil–Mézard conjecture that Henniart first considered the unicity of types in the appendix to [2], where he obtains a positive answer for all representations of $\mathbf{GL}_2$(F). Since then, the result has been extended to cover all supercuspidal representations of $\mathbf{GL}_N$(F) by Paskunas in [15], and to cover all representations of $\mathbf{GL}_3$(F)[14], as well as large classes of representations of $\mathbf{GL}_N$(F)(unpublished at the time of writing) by Nadimpalli.

In this paper, we take the first steps towards unicity for special linear groups, providing a positive answer for supercuspidal representations of $\mathbf{SL}_2(F)$ when $F$ is of odd residual characteristic. The main difference between the cases of $\mathbf{GL}_2(F)$ and $\mathbf{SL}_2(F)$ is that there are now two conjugacy classes of maximal compact subgroups. Given an irreducible representation $\pi$ of $\mathbf{SL}_2(F)$, there will clearly always be a typical representation of at least one of these maximal compact subgroups, obtained by induction from a Bushnell–Kutzko type, but not necessarily on both. In order to make this complication less of a hindrance, we introduce the notion of an *archetype*, which is an $\mathbf{SL}_2(F)$-conjugacy class of types $(K,\tau)$, where K is maximal compact. With this in place, we are able to give a natural extension of the unicity of types to this setting, providing a positive answer for supercuspidal representations in section 2.

In section 3, we go on to provide a more explicit description of the archetypes for the supercuspidal representations of $\mathbf{SL}_2(F)$, obtaining the following result:

**Theorem 1.1.** *Let $\pi$ be a supercuspidal representation of* $\mathbf{SL}_2(F)$, *for F a nonarchimedean local field of odd residual characteristic. If $\pi$ is of integral depth, then $\pi$ contains a unique archetype, while if $\pi$ is of half-integral depth, then there exist precisely two archetypes contained in $\pi$ which are* $\mathbf{GL}_2(F)$*-conjugate but not* $\mathbf{SL}_2(F)$*-conjugate.*

We also provide, in Proposition 3.3, a description of the relationship between supercuspidal archetypes in $\mathbf{GL}_2(F)$ and $\mathbf{SL}_2(F)$ in terms of the local Langlands correspondence, which in some sense says that archetypes are functorial with respect to restriction from $\mathbf{GL}_2(F)$ to $\mathbf{SL}_2(F)$. This allows us to deduce in Corollary 3.4 an extension of Paskunas' inertial Langlands correspondence to our setting.

Our method is to transfer Henniart's results on $\mathbf{GL}_2(F)$ over to $\mathbf{SL}_2(F)$, with the key step being to show that any archetype for an irreducible representation $\bar{\pi}$ of $\mathbf{SL}_2(F)$ must be isomorphic to an irreducible component of the restriction of the unique archetype for some irreducible representation $\pi$ of $\mathbf{GL}_2(F)$ containing $\bar{\pi}$ upon restriction. This is achieved in Lemma 2.3. From this, it is mostly a case of performing simple calculations to deduce in Theorem 2.5 that our unicity result holds. The explicit counting result on the number of archetypes contained in a supercuspidal representation follows easily fromTheorem 2.5, while we are able to prove in Lemma 3.2 a form of converse to Lemma 2.3for the supercuspidal representations, which allows us to easily deduce the remaining results.

The approach taken in this paper is largely general, and already gives partial progress towards a general proof of the unicity of types for $\mathbf{SL}_N(F)$. In particular, the proof ofLemma 2.1 goes through in the general setting without any additional difficulties, suggesting the possibility of applying the results of [15] in a similar manner to our use of Henniart's arguments in order to prove an analogue of Lemma 2.3, which the author is hopeful of managing in the near future. In particular, this would lead easily to a positive answer to the question of unicity. The remaining results should then follow without too much difficulty in a similar manner to that here. In particular, this should allow for the following extension of our explicit results on supercuspidals. Given a supercuspidal representation $\bar{\pi}$ of $\mathbf{SL}_N(F)$ and a supercuspidal representation $\pi$ of $\mathbf{GL}_N(F)$ which contains $\bar{\pi}$ upon restriction, we define the *ramification degree* of $\bar{\pi}$ to be the number $e_{\bar{\pi}}$ such that there are $N/e_{\bar{\pi}}$ characters $\chi$ of $F^\times$ such that $\pi \simeq \pi \otimes (\chi \circ \det)$. This is independent of the choice of $\pi$. Then we make the following conjecture:

**Conjecture 1.2.** *Let* $\bar{\pi}$ *be a supercuspidal representation of* $\mathbf{SL}_N(F)$. *Then there are precisely* $e_{\bar{\pi}}$ *archetypes for* $\bar{\pi}$, *which are* $\mathbf{GL}_N(F)$-*conjugate.*

## 1.1. Notation

Throughout, $F$ will denote a nonarchimedean local field of odd residual characteristic $p$. We will denote by $\mathfrak{O}=\mathfrak{O}_F$ the ring of integers of $F$, and write $\mathfrak{p}=\mathfrak{p}_F$ for its maximal ideal. The residue field will be denoted by $k=k_F=\mathcal{O}/\mathfrak{p}$, and we will write $q$ for the cardinality of $k$. We fix once and for all a choice $\varpi$ of uniformizer of $F$, i.e. an element such that $\varpi\mathcal{O}=\mathfrak{p}$.

When working in generality, we will use G to denote an arbitrary $p$-adic group defined over $F$, by which we will mean the group $G=\mathbf{G}(F)$ of $F$-rational points of some connected reductive algebraic group $\mathbf{G}$ defined over $F$. We will always denote by $G$ the general linear group $\mathbf{GL}_2(F)$. We fix notation for a number of important subgroups of $G$. We will write $K=\mathbf{GL}_2(\mathcal{O})$ for the standard maximal compact subgroup, $T$ for the split maximal torus of diagonal matrices, and $B$ for the standard Borel subgroup of upper triangular matrices. We also write $\bar{G}$ for the special linear group $\mathbf{SL}_2(F)$ and, given a closed subgroup $H$ of $G$, we let $\bar{H}$ denote the subgroup $H\cap\bar{G}$ of $\bar{G}$. We also denote by $T^0$ the compact part of the torus, i.e. the group of diagonal matrices with entries in $\mathcal{O}^\times$, and by $B^0=B\cap K$ the group of upper triangular matrices with entries in $\mathcal{O}$. We will denote by $\eta$ the matrix $\begin{pmatrix} 0 & 1 \\ \varpi & 0 \end{pmatrix}$, so that we may take $\bar{K}$ and $\eta\bar{K}\eta^{-1}$ as representatives of the two $\bar{G}$-conjugacy classes of maximal compact subgroups in $\bar{G}$.

We use the notation ${}^g x = gxg^{-1}$ for conjugation, similarly denoting by ${}^g X = \{{}^g x \mid x\in X\}$ the action of conjugation on a set. Given a representation $\sigma$ of a closed subgroup H of G, we denote by ${}^g\sigma$ the representation of ${}^g\mathcal{H}$ given by ${}^g\sigma(ghg^{-1})=\sigma(h)$.

We write Rep(G) for the category of smooth representations of G, and Irr(G) for the set of isomorphism classes of irreducible representations in Rep(G). Given a closed subgroup H of G, we write $\mathrm{Ind}_{\mathcal{H}}^{\mathcal{G}}\,\sigma$ for the smooth induction of $\sigma$ to G, and $\text{c-Ind}_{\mathcal{H}}^{\mathcal{G}}\,\sigma$ for the compact induction. We write $\mathrm{Res}_{\mathcal{H}}^{\mathcal{G}}\,\pi$ for the restriction of $\pi$ to H, or simply $\pi|_H$ for brevity when it is unnecessary to make clear the functor. Given subgroups $H,H'$ of G and representations $\lambda,\lambda'$ of $H,H'$, respectively, we write $\mathbf{I}_{\mathcal{G}}(\lambda, \lambda') = \{g \in \mathcal{G} \mid \mathrm{Hom}_{\mathcal{H} \cap {}^g\mathcal{H}'}(\lambda, {}^g\lambda') \neq 0\}$ for the intertwining of $\lambda$ with $\lambda'$.

Given a parabolic subgroup P of G with Levi decomposition P=MN, we denote the *normalized* parabolic induction of an irreducible representation $\zeta$ of M to G by $\mathrm{Ind}_{\mathcal{M},\mathcal{P}}^{\mathcal{G}}\ \zeta$. By this, we mean $\mathrm{Ind}_{\mathcal{M},\mathcal{P}}^{\mathcal{G}}\ \zeta = \mathrm{Ind}_{\mathcal{P}}^{\mathcal{G}}\ \tilde{\zeta} \otimes \delta_P^{-1/2}$, where $\tilde{\zeta}$ is the inflation of $\zeta$ to P and $\delta_P$ is the modular character of P.

Finally, we denote by **X**(F) the group of complex characters $\chi : F^\times \to C^\times$. We will be interested in two subgroups of this: the group $\mathbf{X}_{nr}(F)$ of unramified characters in **X**(F) (i.e. those which are trivial on $O^\times$), and the group $\mathbf{X}_N(F)$ of order $N$ characters in **X**(F) (i.e. those $\chi \in \mathbf{X}(F)$ such that $\chi^N = 1$).

## 1.2. The Bernstein decomposition and types

The Bernstein decomposition, which was first introduced in [1], allows us to give a factorization of the category Rep(G), which suggests a natural approach to its study. Given an irreducible representation $\pi$ of G, there exists a unique G-conjugacy class of smooth irreducible representations $\sigma$ of Levi subgroups M of G such that $\pi$ is isomorphic to an irreducible subrepresentation of $\mathrm{Ind}_{\mathcal{H}}^{\mathcal{G}}\ \sigma$, for some parabolic subgroup P of G with Levi factor M. We call this equivalence class the *supercuspidal support* of $\pi$, and denote it by scusp($\pi$). We put a further equivalence relation on the set of possible supercuspidal supports, by saying that (M,$\sigma$) is G-*inertially equivalent* to (M′,$\sigma'$) if there exists a $\chi \in \mathbf{X}_{nr}(F)$ such that (M,$\sigma$) is G-conjugate to (M′,$\sigma' \otimes \chi$). The *inertial support* of $\pi$ is then the inertial equivalence class of scusp($\pi$). If scusp($\pi$)=(M,$\sigma$), then we write $[M,\sigma]_G$ for the inertial support of $\pi$.

With this in place, let B(G) denote the set of inertial equivalence classes of supercuspidal supports, and, for $s \in B(G)$, let $\mathrm{Rep}^s(G)$ denote the full subcategory of Rep(G) consisting of representations such that all irreducible subquotients have inertial support s, and write $\mathrm{Irr}^{s}(G)$ for the set of isomorphism classes of irreducible representations in $\mathrm{Rep}^{s}(G)$. Bernstein then shows that

$$\mathrm{Rep}(\mathcal{G}) = \prod_{\mathfrak{s}\in\mathfrak{B}(\mathcal{G})} \mathrm{Rep}^{\mathfrak{s}}(\mathcal{G}).$$

More generally, given a subset $\mathfrak{S}$ of $\mathfrak{B}(\mathcal{G})$, let $\mathrm{Rep}^{\mathfrak{S}}(\mathcal{G}) = \prod_{\mathfrak{s}\in\mathfrak{S}} \mathrm{Rep}^{\mathfrak{s}}(\mathcal{G})$ and $\mathrm{Irr}^{\mathfrak{S}}(\mathcal{G}) = \bigcup_{\mathfrak{s}\in\mathfrak{S}} \mathrm{Irr}^{\mathfrak{s}}(\mathcal{G})$. This allows us to define the notion of a type in generality:

**Definition 1.3.**

Let $S \subset B(G)$. Let $(J,\lambda)$ be a pair consisting of a compact open subgroup $J$ of G and a smooth irreducible representation $\lambda$ of $J$.

(i)We say that $(J,\lambda)$ is S-*typical* if, for any smooth irreducible representation $\pi$ of G, we have that $\mathrm{Hom}_J(\pi\restriction_J, \lambda) \neq 0 \Rightarrow \pi \in \mathrm{Irr}^{\mathfrak{S}}(\mathcal{G})$.

(ii)We say that $(J,\lambda)$ is an S-*type* if it is S-typical, and $\mathrm{Hom}_J(\pi\restriction_J, \lambda) \neq 0$ for each $\pi \in \mathrm{Irr}^{S}(G)$.

In the case that S={s} is a singleton, we will simply speak of s-types rather than {s}-types.

In the cases of interest to us, the Bernstein components of Rep(G) admit particularly simple descriptions: if $\mathrm{Rep}^{s}(G)$ contains a supercuspidal representation $\pi$, then $\mathrm{Irr}^{s}(G) = \{\pi\otimes(\chi\circ\det) \mid \chi\in\mathbf{X}_{nr}(F)\}$. The situation for $\bar{G}$ is even simpler: as $\bar{G}$ has no unramified characters, $\mathrm{Irr}^{\mathfrak{s}}(\bar{G})$ is a singleton whenever it contains a supercuspidal representation.

We now introduce the slightly modified notion of an *archetype*, which is more suited to studying the unicity of types in groups other than $\mathbf{GL}_N(F)$.

**Definition 1.4.** Let $\mathfrak{S} \subset \mathfrak{B}(\mathcal{G})$. An S-*archetype* is a G-conjugacy class of S-typical pairs $(\mathscr{K}, \tau)$ for $\mathscr{K}$ a maximal compact subgroup of G. Given a representative (K,τ) of an archetype, we write ${}^{\mathcal{G}}(\mathscr{K}, \tau)$ for the full conjugacy class.

**Remark 1.5.**

It may seem odd to define an archetype as a conjugacy class of *typical* representations rather than as a conjugacy class of types. However, for us, the difference turns out to be unimportant: the unicity of types will allow us to see that typical representations of maximal compact subgroups are types in almost all cases (indeed, for all representations not contained in the restriction of the Steinberg representation of *G* ). The reason for working with typical representations rather than types is that it allows us to include these "Steinberg" representations in the general picture, despite them admitting no type of the form (K,τ).

There is one obvious way of constructing archetypes:

**Lemma 1.6.** *Let π be an irreducible representation of a p-adic group* G*of inertial support* s. *Let* (J,λ)*be an* s-*type, and let* K*be a maximal compact subgroup of* G*containing J. Then the irreducible components of* $\tau := c\text{-}\mathrm{Ind}_J^{\mathscr{K}}$ *are representatives of* s-*archetypes. Moreover, if τ is irreducible then it is an* s-*type.*

**Proof.** Using Frobenius reciprocity, it is clear that if an irreducible representation π′ of Gcontains *τ*, then it must contain *λ*, hence the first claim. The second claim simply follows by the transitivity of induction. □

The question of the unicity of types is then whether there are any archetypes other than those induced from Bushnell–Kutzko types. For *G*, Henniart answers this in the appendix to [2]:

**Theorem 1.7.** *Suppose* q≠2. *Let π be an irreducible representation of G of inertial support* s. *Let* ${}^{\mathcal{G}}(\mathscr{K}, \tau)$*be an* s-*archetype. Then there* *exists a Bushnell–Kutzko type* (J,λ)*with* J⊂K*such that* $\tau \hookrightarrow c\text{-Ind}_J^K \lambda$. *Moreover, unless π is a twist of the Steinberg representation* $St_G$, *the representation* $c\text{-Ind}_J^K \lambda$ *is irreducible and hence* (K,τ)*is an* s-*type.*

## 1.3. Bushnell–Kutzko types

We now describe the explicit construction, due to Bushnell and Kutzko, of types for the irreducible representations of $G$ and $\bar{G}$. In this section, we discuss the types for supercuspidal representations, which are the *simple types* constructed in [4], [5] and [6]. The construction of these types is by a series of successively stronger approximations of a type, and is rather technical in nature. We omit as many details as possible; the full details for our case of N=2 may be found in the appendix of [2], or in [3]. The starting points for the construction are the *hereditary* O-*orders.* For our purposes, we may simply say that the $G$ -conjugacy classes of hereditary orders in $Mat_2(F)$ are represented by the maximal order $M=Mat_2(O)$, and the Iwahori order I, which consists of those matrices in M which are upper-triangular modulo p. The parahoric subgroups of $G$ are then the groups of units of these rings. Letting $U_M=M^\times$ and $U_I=I^\times$, we may take as representatives for the $\bar{G}$-conjugacy classes of parahoric subgroups of $\bar{G}$ the groups $\bar{U}_{\mathfrak{M}} = U_{\mathfrak{M}} \cap \bar{G}$, its conjugate ${}^{\eta}\bar{U}_{\mathfrak{M}}$, and $\bar{U}_{\mathfrak{I}} = U_{\mathfrak{I}} \cap \bar{G}$.

We also require the Jacobson radicals of these hereditary orders. The radical of M is $P_M=Mat_2(p)$, and the radical of I is the ideal $P_I$ of matrices which are strictly upper-triangular modulo p. Given a hereditary order A, we may then define a filtration of $U_A$ by compact open subgroups, by setting $U_{\mathfrak{A}}^n = 1+\mathfrak{P}_{\mathfrak{A}}^n$, for $n \geq 1$. There is an integer $e_A$ called the O-*lattice period* associated to each hereditary order; it is the positive integer $e_A$ such that $\mathfrak{P}_{\mathfrak{A}}^{e_{\mathfrak{A}}} = \varpi\mathfrak{A}$. The construction of the simple types (J,λ) is then by *simple strata* . Roughly speaking, any type (J,λ) for a supercuspidal representation π of $G$ is constructed via a triple [A,n,β] consisting of a hereditary O-order A, the integer $n$ such that $n/e_A$ is the depth of $\pi$, and an element $\beta$ of $\mathfrak{P}_{\mathfrak{A}}^{-n}$ such that E:=F[β] is a field. For our purposes, it suffices to know that $J = \mathfrak{o}_E^{\times} U_{\mathfrak{A}}^{\lfloor \frac{n+1}{2} \rfloor}$. We will also briefly make use of certain filtration subgroups of $J$: for an integer k≥1, let $J^k = J \cap U_{\mathfrak{A}}^k$.

These constructions lead, for each supercuspidal representation $\pi$ of $G$, to an irreducible representation $\lambda$ of a compact open subgroup $J$ of $G$ , such that (J,λ) is a $[G,\pi]_G$-type and there exists a unique extension $\Lambda$ of $\lambda$ to the $G$ -normalizer $\tilde{J}$ of $J$ such that $\pi \simeq c\text{-}\mathrm{Ind}_{\tilde{J}}^{G}$.A Any s-type arising from these constructions is a *(maximal) G-simple type*. The other main fact that we will require is the "intertwining implies conjugacy" property [4, Theorem 5.7.1], which says that, if we have two maximal simple types (J,λ)and (J′,λ′) such that $\mathbf{I}_G(\lambda,\lambda')\neq\emptyset$, then (J,λ) and (J′,λ′) must actually be $G$-conjugate.

In our case, the simple types in $\bar{G}$ are easily obtained from those in $G$. Let $\pi$ be a supercuspidal representation of , so that $\pi \downharpoonright_{\bar{G}}$ splits into a finite sum of supercuspidal representations of $\bar{G}$. Choose a simple type (J,λ) extending to $(\tilde{J}, \Lambda)$ such that $\pi \simeq c\text{-}\mathrm{Ind}_{\tilde{J}}^{G}$, so that we may perform a Mackey decomposition to obtain

$$\pi \downharpoonright_{\bar{G}} \simeq \bigoplus_{\bar{G}\backslash G/\tilde{J}} c\text{-}\mathrm{Ind}_{{}^g\bar{J}}^{\bar{G}}\ {}^g\bar{\lambda},$$

where $\bar{\lambda} = \lambda \downharpoonright_{\bar{J}}$. This is a finite length sum, and the summands will generally be reducible of finite length. However, in our case all ramification is tame and this is actually a decomposition into irreducibles, with one family of exceptions: for the unramified twists of the "exceptional depth zero" supercuspidal representation of $G$, which under local Langlands corresponds to the triple imprimitive representation of the Weil group, each of the above summands is reducible of length 2 – see [6] for details. We then define the *(maximal)* $\bar{G}$*-simple types* to be the irreducible components of the representations ${}^g\bar{\lambda}$, for (J,λ) running over the $G$ -simple types.

Given such a $\bar{G}$-simple type $(\bar{J}, \mu)$, we have that $\mathrm{I}_{\bar{G}}(\mu) = \bar{J}$; thus they induce up to a supercuspidal representation of $\bar{G}$, and it is clear that this gives a construction of all of the supercuspidals of $\bar{G}$. Just as in the case of $G$, we have an intertwining implies conjugacy property: if two maximal $\bar{G}$-simple types $(\bar{J}, \mu)$ and $(\bar{J}', \mu')$ are such that $\mathrm{I}_{\bar{G}}(\mu, \mu') \neq \emptyset$, then there exists a $g \in \bar{G}$ such that $(\bar{J}', \mu') \simeq ({}^g\bar{J}, {}^g\mu)$[4, Theorem 5.3 and Corollary 5.4].

## 1.4. The local Langlands correspondence for supercuspidals

Some of our results on supercuspidals will require a basic understanding of the relevant local Langlands correspondences, which we quickly recall here. Fix once and for all a choice $\bar{F}/F$ of separable algebraic closure. Let $I_F$ denote the inertia subgroup of $\mathrm{Gal}(\bar{F}/F)$, i.e. the kernel of the natural projection $\mathrm{Gal}(\bar{F}/F) \twoheadrightarrow \mathrm{Gal}(\bar{\mathfrak{k}}/\mathfrak{k})$, and let $W_F$ denote the Weil group of $\bar{F}/F$, i.e. the subgroup of $\mathrm{Gal}(\bar{F}/F)$ generated by $I_F$ and the Frobenius elements, topologized so that $I_F$ is open in $W_F$ and given the subspace topology from $\mathrm{Gal}(\bar{F}/F)$.

For a $p$-adic group G, let $\mathrm{Irr}_{\mathrm{scusp}}(G)$ denote the set of equivalence classes of supercuspidal representations of G. Let $L^0(G)$ denote the set of irreducible $L$-parameters for G, which is the same as the set of irreducible Frobenius-semisimple representations $W_F \to \mathbf{GL}_2(\mathbb{C})$, i.e. those irreducible representations under which some fixed Frobenius element of $W_F$ acts semisimply. Then the local Langlands correspondence for $G$ provides a unique natural bijection rec: $\mathrm{Irr}_{\mathrm{scusp}}(G) \leftrightarrow \mathcal{L}^0(G)$, which preserves $L$-functions and $\varepsilon$-factors, as well as mapping supercuspidal representations to irreducible $L$-parameters, among a list of other properties.

From this, as shown in [12] and [9], one may deduce a Langlands correspondence for the supercuspidal representations of $\bar{G}$, which suffices for our purposes. Let $\mathcal{L}^0(\mathcal{G})$ be the image of $L^0(G)$ under the natural map $\mathrm{Hom}(W_F,\mathbf{GL}_2(C))\to\mathrm{Hom}(W_F,\mathbf{PGL}_2(C))$. Then we define the local Langlands correspondence $\overline{\mathrm{rec}}$ on $\mathrm{Irr}_{\mathrm{scusp}}(\bar{G})$ by requiring that, for *any* map $R$ which sends a supercuspidal representation $\pi$ of $G$ to one of the irreducible components of $\mathrm{Res}_{\bar{G}}^{G}\ \pi$, the diagram

$$\begin{array}{ccc} \mathrm{Irr}_{\mathrm{scusp}}(G) & \xleftrightarrow{\mathrm{rec}} & \mathcal{L}^0(G) \\ {\scriptstyle R}\downarrow & & \downarrow \\ \mathrm{Irr}_{\mathrm{scusp}}(\bar{G}) & \xrightarrow{\overline{\mathrm{rec}}} & \mathcal{L}^0(\bar{G}) \end{array}$$

commutes. The (supercuspidal) $L$ -packets are then simply the finite fibres of the map $\overline{\mathrm{rec}}$, which are precisely the sets of irreducible components of the restrictions to $\bar{G}$ of supercuspidal representations of $G$.

One may use the local Langlands correspondence to give an alternative description of $G$-inertial equivalence classes of supercuspidal representations: two supercuspidal representations $\pi$ and $\pi'$ of $G$ are inertially equivalent if and only if $\mathrm{rec}(\pi)\restriction_{I_F}\simeq\mathrm{rec}(\pi')\restriction_{I_F}$.

## 2. THE MAIN UNICITY RESULT

We now begin working towards the main results, beginning with a description of the relationship between archetypes in $G$ and those in $\bar{G}$.

**Lemma 2.1.**

*Let $\pi$ be a supercuspidal representation of G, let $\bar{\pi}$ be an irreducible component of $\pi\restriction_{\bar{G}}$, and suppose that $\bar{\pi}$ admits an archetype ${}^{\bar{G}}(\bar{K},\bar{\tau})$. Let $\Psi$ be an irreducible subquotient of $\mathrm{Ind}_{\bar{K}}^{K}\ \bar{\tau}$ which is contained in $\pi\restriction_K$, and let $S=\{[G,\pi\otimes(\chi\circ\det)]_G \mid \chi\in\mathbf{X}_2(F)\}$. Then $\Psi$ is S-typical.*

**Proof.**

We first note that such a $\Psi$ clearly exists: let $\omega_\pi$ denote the central character of $\pi$ , and write $\omega_\pi^0$ for its restriction to $O^\times$. Let $\tilde{\tau}$ be the extension to $\mathcal{O}^\times \bar{K}$ of $\bar{\tau}$ by $\omega_\pi^0$. Then, by Frobenius reciprocity, some irreducible quotient of $c\text{-}\mathrm{Ind}_{\mathcal{O}^\times \bar{K}}^{K} \tilde{\tau}$ must be contained in $\pi$ upon restriction to $K$. From now on, $\Psi$ will always denote this representation.

Let $\pi'$ be an irreducible representation of $G$, and suppose that $\mathrm{Hom}_K(\pi' \downharpoonright_K, \Psi) \neq 0$. Then

$$\begin{aligned} 0 &\neq \mathrm{Hom}_K(\mathrm{Ind}_{\bar{K}}^{K}\ \bar{\tau}, \mathrm{Res}_K^G\ \pi') \\ &= \mathrm{Hom}_{\bar{K}}(\bar{\tau}, \mathrm{Res}_{\bar{K}}^{\bar{G}} \mathrm{Res}_{\bar{G}}^{G}\ \pi'). \end{aligned}$$

Since $\bar{G}(\bar{K}, \bar{\tau})$ is an archetype for $\bar{\pi}$, we see that $\pi'$ must contain $\bar{\pi}$ upon restriction to $\bar{G}$, so that $\pi'$ is of inertial support $[G, \pi\otimes(\chi\circ\det)]_G$, for some $\chi\in\mathbf{X}(F)$. Comparing central characters, $\chi$ must be an unramified twist of a quadratic character, as required.

**Remark 2.2.**

Of course, the same result holds true with an identical proof when $\bar{\pi}$ admits an archetype of the form $\bar{G}({}^{\eta}\bar{K}, \bar{\tau})$; i.e. if $\Psi$ is an irreducible subquotient of $\mathrm{Ind}_{{}^{\eta}\bar{K}}^{{}^{\eta}K}\ \bar{\tau}$ which is contained in $\pi \downharpoonright_{{}^{\eta}\bar{K}}$, then $\Psi$ is S-typical.

**Lemma 2.3.**

*The representation $\Psi$ constructed in Lemma 2.1is a $[G,\pi]_G$-type.*

**Proof.**

By Lemma 2.1, it remains only to rule out the possibility that $\Psi$ is contained in a representation of inertial support $[G,\pi\otimes(\chi\circ\det)]_G$, for some non-trivial $\chi\in\mathbf{X}_2(F)$. Indeed, this would show that $\Psi$ is $[G,\pi]_G$-typical, and the unicity of types for $G$ would immediately imply that $\Psi$ represents a $[G,\pi]_G$-archetype. We now argue by cases.

As $\pi$ is a supercuspidal representation, we may write $\pi \simeq c\text{-}\mathrm{Ind}_{\tilde{J}}^{G}\ \Lambda$, with $(\tilde{J},\Lambda)$ extending a maximal simple type $(J,\lambda)$ contained in $\pi$. Suppose for contradiction that $\Psi$ is not a type. As noted by Henniart in the appendix to [2, paragraphs A.2.4–A.2.7, A.3.6–A.3.7 and A.3.9–A.3.11], every irreducible component of $\pi\downharpoonright_K$ other than the typical representation $\tau = \mathrm{Ind}_J^K\ \lambda$ appears in the restriction to $K$ of either a parabolically induced representation, or some other supercuspidal $\pi_\mu$, in a different inertial equivalence class to that of $\pi$, which we may now describe explicitly. There are three further subcases which we treat separately.

Suppose first that $\Psi$ is contained in some parabolically induced representation. We may therefore find a character $\zeta$ of $T$ such that $\Psi$ is isomorphic to some irreducible component of $\mathrm{Res}_K^G\,\mathrm{Ind}_{T,B}^G\ \zeta$. As $\mathrm{Ind}_{\bar{K}}^{K}\ \bar{\tau}$ projects onto $\Psi$, we therefore have

$$
\begin{aligned}
0 &\neq \mathrm{Hom}_K(c\text{-}\mathrm{Ind}_{\bar{K}}^{K}\ \bar{\tau}, \mathrm{Res}_K^G\,\mathrm{Ind}_{T,B}^G\ \zeta)\\
&= \mathrm{Hom}_{\bar{G}}(c\text{-}\mathrm{Ind}_{\bar{K}}^{G}\ \bar{\tau}, \mathrm{Res}_{\bar{G}}^{G}\,\mathrm{Ind}_{T,B}^G\ \zeta)\\
&= \bigoplus_{i=1}^{n} \mathrm{Hom}_{\bar{G}}(\bar{\pi}, \mathrm{Ind}_{\bar{T},\bar{B}}^{\bar{G}}\,\mathrm{Res}_{\bar{T}}^{T}\ \zeta).
\end{aligned}
$$

Here, $n$ is the integer such that $c\text{-}\mathrm{Ind}_{\bar{K}}^{\bar{G}}\ \bar{\tau} \simeq \bar{\pi}^{\oplus n}$, which exists by Proposition 5.2 of[7], and the final equality follows from a Mackey decomposition with the summation involved being trivial as $B\bar{G} = G$. Hence $\bar{\pi}$ is contained in some parabolically induced representation, which provides a contradiction by Lemma 2.1.

Now suppose $\Psi$ does not appear as an irreducible component of the restriction to $K$ of any parabolically induced representation, and suppose furthermore that $\pi$ is of integral depth $n$. In this case, we may construct a new supercuspidal representation containing every irreducible component of $\pi\lfloor_K$ other than the archetype $\tau$. Let E/F be the unique unramified quadratic extension of $F$, and choose an embedding $\mathcal{O}_E^\times \subset K$. Let $\mu$ be any level 1 character of $E^\times$ trivial on $F^\times$, and let $(J,\lambda)$ be a simple type for $\pi$. Then the pair $(J,\lambda\otimes\mu)$ is again a maximal simple type contained in some supercuspidal representation $\pi_\mu$ lying in a different inertial equivalence class to that of $\pi$, and any irreducible component of $\pi\lfloor_K$ other than $\tau$ must be contained in $\pi_\mu$ upon restriction; in particular, we must have $\Psi\hookrightarrow\pi_\mu\lfloor_K$. But then $\pi_\mu$ must be isomorphic to an unramified twist of $\pi\otimes(\chi\circ\det)$, for some (non-trivial by assumption) $\chi\in\mathbf{X}_2(F)$, which is to say that their archetypes must coincide. The archetype for $\pi_\mu$ is $\text{c-Ind}_J^K\ \lambda\otimes\mu$, and the archetype for $\pi\otimes(\chi\circ\det)$ is $(\text{c-Ind}_J^K\ \lambda)\otimes(\chi\circ\det)$. If these two representations are isomorphic, then we must have $\lambda\otimes\mu\simeq\lambda\otimes(\chi\circ\det)$, as $\mathbf{I}_K(\lambda\otimes\mu,\lambda\otimes(\chi\circ\det))\neq\emptyset$, and if $g$ intertwines $\lambda\otimes\mu$ with $\lambda\otimes(\chi\circ\det)$, then $g$ intertwines $\lambda\lfloor_{J^1}$ with itself, and so $g\in J$. As $\lambda\otimes\mu\simeq\lambda\otimes(\chi\circ\det)$, we may use Schur's lemma to obtain $0\neq\mathrm{Hom}_J(\lambda\otimes\mu,\lambda\otimes(\chi\circ\det))\subseteq\mathrm{End}_{J^1}(\lambda\lfloor_{J^1})=\mathbb{C}$, and hence $\mathrm{Hom}_J(\lambda\otimes\mu,\lambda\otimes(\chi\circ\det))$ contains the identity map, so that we must have $\mu=\chi\circ\det$ on $\mathcal{O}_E^\times$. However, there are only two quadratic characters $\chi$ of $F^\times$ while there are $q+1\geq4$ such characters $\mu$. Choosing $\mu$ non-quadratic, we obtain a contradiction.

Finally, consider the case where $\pi$ is of half-integral depth and $\Psi$ is not contained in any parabolically induced representation, we argue essentially as before. We assume further that $\pi$ is of depth at least $\frac{3}{2}$; for $\pi$ of depth $\frac{1}{2}$ Henniart shows that this case never arises. Let E/F be the ramified quadratic extension associated to the simple type for $\pi$, and choose an embedding $\mathcal{O}_E^\times\subset U_{\mathfrak{J}}\subset K$. For $\mu$, we take a level 2 character of $E^\times$ trivial on $\mathcal{O}_F^\times$, and construct $\pi_\mu$ as before. Letting $(J,\lambda)$ be a simple type for $\pi$, so that $\lambda$ is one-dimensional, the pair $(J,\lambda\otimes\mu)$ is a simple type for some supercuspidal $\pi_\mu$ in a different inertial equivalence class to that of $\pi$, and $\Psi$ must appear in the restriction to $K$ of $\pi_\mu$. Then, up to an unramified twist, $\pi_\mu\simeq\pi\otimes(\chi\circ\det)$ for $\chi$ a nontrivial quadratic character of $F^\times$, and so the archetypes $c\text{-}\mathrm{Ind}_J^K\ \lambda\otimes\mu$ and $(c\text{-}\mathrm{Ind}_J^K\ \lambda)\otimes(\chi\circ\det)$ coincide. Then $\mathbf{I}_K(\lambda\otimes\mu,\lambda\otimes(\chi\circ\det))\neq\emptyset$, and if $g\in\mathbf{I}_K(\lambda\otimes\mu,\lambda\otimes(\chi\ \det))$, then $g\in\mathbf{I}_K(\lambda\restriction_{J^2},\lambda\restriction_{J^2})=J$, as $\pi$ is of depth at least $\frac{3}{2}$, and $\lambda$ is one-dimensional so that $\lambda\restriction_{J^2}$ is a simple character. It follows that we must have $\lambda\otimes\mu\simeq\lambda\otimes(\chi\circ\det)$, and so $\mu\simeq\chi\circ\det$. But then $\mu$ is of level 2 while $\chi\circ\det$ is tame and hence of level at most 1, giving a contradiction and completing the proof. □

**Lemma 2.4.** *Let $\pi$ be a supercuspidal representation of G, and let ${}^G(K,\tau)$ be an archetype for $\pi$. Then every irreducible component of $\tau\restriction_{\bar K}$ is induced from a Bushnell–Kutzko type for $\bar G$.*

**Proof.** By Theorem 1.7, the representation $\tau$ is of the form $\tau=c\text{-}\mathrm{Ind}_J^K\ \lambda$, where $(J,\lambda)$ is a maximal simple type contained in $\pi$. Then we may perform a Mackey decomposition to obtain

$$\mathrm{Res}_{\bar K}^K\, c\text{-}\mathrm{Ind}_J^K\ \lambda=\bigoplus_{J\backslash K/\bar K} c\text{-}\mathrm{Ind}_{{}^g\bar J}^{\bar K}\ {}^g\bar\lambda,$$

where $\bar\lambda=\lambda\restriction_{\bar J}$. The irreducible components of this representation are all of the required form. □

**Theorem 2.5.** *Let $\bar\pi$ be a supercuspidal representation of $\bar G$*

*(i)If ${}^{\bar G}(\mathscr{K},\bar\tau)$ is an archetype for $\bar\pi$, then there exists a simple type $(\bar J,\mu)$ with $\bar J\subset\mathscr{K}$ such that $\bar\tau\simeq c\text{-}\mathrm{Ind}_{\bar J}^{\mathscr{K}}\ \mu$.*

*(ii)If $(\bar J,\mu)$ is a simple type contained in $\bar\pi$ and* K*is a maximal compact subgroup of $\bar G$ which contains $\bar J$, then the representation $c\text{-}\mathrm{Ind}_{\bar J}^{\mathscr{K}}\ \mu$ is the unique $[\bar G,\bar\pi]_{\bar G}$-typical representation of* K.

**Proof.**For *(i)* we may reduce to the case $\mathscr{K}=\bar K$. By Lemma 2.3, $\bar\tau$ is an irreducible component of the restriction to $\bar K$ of the unique archetype (K,τ) for some irreducible representation $\pi$ of $G$ containing $\bar\pi$ upon restriction to $\bar G$. As there exists a simple type(J,λ) such that $\tau = c\text{-}\operatorname{Ind}_J^K \lambda$, the result follows immediately from Lemma 2.4.

To see *(ii)* , it remains to check that, given two distinct simple types $(\bar{J}, \mu)$ and $(\bar{J}', \mu')$ contained in $\bar{\pi}$ which are, moreover, contained in the same conjugacy class of maximal compact subgroups, these simple types provide the same archetypes through induction. Thus, we may as well assume that $\bar{J}, \bar{J}' \subset \bar{K}$. As $(\bar{J}, \mu)$ and $(\bar{J}', \mu')$ are $[\bar{G}, \bar{\pi}]_{\bar{G}}$-types, $\bar{\pi}$ will appear as a subquotient of the induced representations $c\text{-}\operatorname{Ind}_{\bar{J}}^{\bar{G}} \mu$ and $c\text{-}\operatorname{Ind}_{\bar{J}'}^{\bar{G}} \mu'$; hence we will have

$$
\begin{aligned}
0 &\neq \operatorname{Hom}_{\bar{G}}(c\text{-}\operatorname{Ind}_{\bar{J}}^{\bar{G}} \mu, c\text{-}\operatorname{Ind}_{\bar{J}'}^{\bar{G}} \mu') \\
&= \operatorname{Hom}_{\bar{J}}(\mu, \operatorname{Res}_{\bar{J}}^{\bar{G}} c\text{-}\operatorname{Ind}_{\bar{J}'}^{\bar{G}} \mu') \\
&= \bigoplus_{\bar{J}'\backslash\bar{G}/\bar{J}} \operatorname{Hom}_{\bar{J}}(\mu, c\text{-}\operatorname{Ind}_{{}^g\bar{J}'\cap\bar{J}}^{\bar{J}} \operatorname{Res}_{{}^g\bar{J}'\cap\bar{J}}^{{}^g\bar{J}'} {}^g\mu') \\
&= \bigoplus_{\bar{J}'\backslash\bar{G}/\bar{J}} \operatorname{Hom}_{{}^g\bar{J}'\cap\bar{J}}(\operatorname{Res}_{{}^g\bar{J}'\cap\bar{J}}^{\bar{J}} \mu, \operatorname{Res}_{{}^g\bar{J}'\cap\bar{J}}^{{}^g\bar{J}'} {}^g\mu'),
\end{aligned}
$$

and so $I_{\bar{G}}(\mu, \mu') \neq \emptyset$. As $\bar{\pi}$ is supercuspidal, $(\bar{J}, \mu)$ and $(\bar{J}', \mu')$ will be simple types, and so by the intertwining implies conjugacy property there will exist a $g \in \bar{G}$ such that ${}^g(c\text{-}\operatorname{Ind}_{\bar{J}}^{\bar{K}} \mu) \simeq c\text{-}\operatorname{Ind}_{\bar{J}'}^{{}^g\bar{K}} \mu'$. As $\bar{J}'$ is contained in at most one maximal compact subgroup in each $\bar{G}$-conjugacy class, we must actually have ${}^g\bar{K} = \bar{K}$, and so $(\bar{J}, \mu)$and $(\bar{J}', \mu')$ induce to the same archetype. □

# 3. AN EXPLICIT DESCRIPTION OF SUPERCUSPIDAL ARCHETYPES AND A RESULT ON *L*-PACKETS

Having completed the proof of Theorem 2.5, we now provide a more explicit description of the theory of archetypes for supercuspidal representations. Given a supercuspidal representation $\bar{\pi}$ of $\bar{G}$, we define the *ramification degree* $e_{\bar{\pi}}$ of $\bar{\pi}$ to be 1 if $\bar{\pi}$ is of integral depth, or 2 if $\bar{\pi}$ is of half-integral depth. Then we obtain the following corollary to Theorem 2.5:

**Corollary 3.1.***Let $\bar{\pi}$ be a supercuspidal representation of $\bar{G}$. Then the number of $[\bar{G}, \bar{\pi}]_{\bar{G}}$-archetypes is precisely $e_{\bar{\pi}}$.*

**Proof.**It remains only for us to count the number of archetypes obtained by inducing a maximal simple type contained in $\bar{\pi}$ up to maximal compact subgroups. If $\bar{\pi}$ is ramified then, up to conjugacy, any simple type for $\bar{\pi}$ is defined on a group contained in the Iwahori subgroup $\bar{U}_{\mathfrak{J}}$ of $\bar{G}$, which is itself contained in both $\bar{K}$ and ${}^{\eta}\bar{K}$; hence ramified supercuspidals admit two archetypes. If $\bar{\pi}$ is unramified, it suffices to show that the subgroup $\bar{J}$ on which any simple type $\mu$ for $\bar{\pi}$ is defined embeds into precisely one $\bar{G}$-conjugacy class of maximal compact subgroups. Without loss of generality, we may as well assume that $\bar{J} \subseteq \bar{U}_{\mathfrak{M}}$. We have $\ker \mathrm{N}_{E/F} \subseteq \bar{J} \subseteq \bar{U}_{\mathfrak{M}}$, where $\mathbf{N}_{E/F}$ is the norm map on the (unramified) quadratic extension E/F associated to $\bar{\pi}$. Suppose for contradiction that we also have $\bar{J} \subseteq {}^{\eta}\bar{U}_{\mathfrak{M}}$. As the group $\ker \mathbf{N}_{E/F}$ contains the group $\mu_{q+1}$ of (q+1)-th roots of unity, we would therefore also have $\mu_{q+1} \subset \bar{U}_{\mathfrak{M}} \cap {}^{\eta}\bar{U}_{\mathfrak{M}} = \bar{U}_{\mathfrak{J}}$. However, the Iwahori subgroup contains no order q+1 elements, giving the desired contradiction.

Thus, the only way in which one might obtain two archetypes when $e_{\bar{\pi}} = 1$ is if $\bar{\pi}$ contains simple types which are $G$ -conjugate but not $\bar{G}$-conjugate; this clearly cannot be the case by the intertwining implies conjugacy property. □

This completely describes the number of archetypes contained in any supercuspidal representation of $\bar{G}$. We now prove a complementary result, which allows us to describe the relationship between the theories of archetypes for $\bar{G}$ and $G$. We first require a converse result to Lemma 2.3.

**Lemma 3.2.**

*Let π be a supercuspidal representation of G, let* s=[G,π]$_G$, *and let* ${}^{G}(K,\tau)$ *be the unique* s*-archetype. Let* $\bar{\pi}$ *be an irreducible component of* $\pi \downarrow_{\bar{G}}$. *Then there exists a* g∈G*and an irreducible component* $\bar{\tau}$ *of* ${}^{g}\tau \downarrow_{{}^{g}\bar{K}}$ *such that* ${}^{\bar{G}}({}^{g}\bar{K},\bar{\tau})$ *is an archetype for* $\bar{\pi}$.

**Proof.**

We may assume without loss of generality, by conjugating by $\eta$ if necessary, that $\bar{\pi} = c\text{-}\mathrm{Ind}_{\bar{K}}^{\bar{G}}\ \rho$, where $\rho = c\text{-}\mathrm{Ind}_{\bar{J}}^{\bar{K}}\ \mu$ is the induction to $\bar{K}$ of a $\bar{G}$-simple type. Let $\{\bar{\tau}_j\}$ be the (finite) set of irreducible components of $\tau \downarrow_{\bar{K}}$. We first show that any $\pi' \in \mathrm{Irr}(\bar{G})$ containing one of the $\bar{\tau}_j$ upon restriction must appear in the restriction to $\bar{G}$ of $\pi$. We have

$$0 \neq \bigoplus_j \mathrm{Hom}_{\bar{K}}(\bar{\tau}_j, \pi')$$

$$= \mathrm{Hom}_{\bar{K}}(\mathrm{Res}_{\bar{K}}^{K}\ \tau, \mathrm{Res}_{\bar{K}}^{\bar{G}}\ \pi')$$

$$= \mathrm{Hom}_{\bar{G}}(c\text{-}\mathrm{Ind}_{\bar{K}}^{\bar{G}}\ \mathrm{Res}_{\bar{K}}^{K}\ \tau, \pi'),$$

and so we obtain $\pi' \twoheadleftarrow c\text{-}\mathrm{Ind}_{\bar{K}}^{\bar{G}}\ \mathrm{Res}_{\bar{K}}^{K}\ \tau \hookrightarrow \mathrm{Res}_{\bar{G}}^{G}\ c\text{-}\mathrm{Ind}_{K}^{G}\ \tau$. Every irreducible subquotient of the representation $c\text{-}\mathrm{Ind}_{K}^{G}\ \tau$ is a twist of $\pi$, and hence coincides with $\pi$ upon restriction to $\bar{G}$, so that any such representation π′ must be a subrepresentation of the restriction to $\bar{G}$ of $\pi$ . Hence the possible representations π′all lie in a single $G$ -conjugacy class of representations of $\bar{G}$. Let g∈G be such that ${}^{g}\pi' \simeq \bar{\pi}$, so that $\pi' \simeq c\text{-}\mathrm{Ind}_{{}^{g}\bar{K}}^{\bar{G}}\ {}^{g}\rho$, and choose $j$ so that π′ contains $\bar{\tau}_j$. We claim that $({}^{g}\bar{K}, {}^{g}\bar{\tau}_j)$ is the required type.

It suffices to show that any $G$ -conjugate of $\bar{\pi}$ containing $({}^{g}\bar{K}, {}^{g}\bar{\tau}_j)$ is isomorphic to $\bar{\pi}$. Suppose that, for some h∈G, we have

$\mathrm{Hom}_{^g\bar{K}}({}^h\bar{\pi}, {}^g\bar{\tau}_j) \neq 0$. The representation ${}^h\bar{\pi}$ is of the form ${}^h\bar{\pi} = c\text{-}\mathrm{Ind}_{^h\bar{J}}^{\bar{G}}\ {}^h\mu$ and, using Lemma 2.4, we see that the representation ${}^g\bar{\tau}_j$ must be induced from some $\bar{G}$-simple type $(\bar{J}', \mu')$, say. Then

$$\begin{aligned}
0 &\neq \mathrm{Hom}_{^g\bar{K}}(\mathrm{Res}_{^g\bar{K}}^{\bar{G}}\ {}^h\bar{\pi}, {}^g\bar{\tau}_j) \\
&= \mathrm{Hom}_{\bar{J}'}(\mathrm{Res}_{\bar{J}'}^{\bar{G}}\ c\text{-}\mathrm{Ind}_{^h\bar{J}}^{\bar{G}}\ {}^h\mu, \mu') \\
&= \bigoplus_{^h\bar{J}\backslash\bar{G}/\bar{J}'} \mathrm{Hom}_{\bar{J}'}(c\text{-}\mathrm{Ind}_{^{xh}\bar{J}\cap\bar{J}'}^{\bar{J}'}\ \mathrm{Res}_{^{xh}\bar{J}\cap\bar{J}'}^{^{xh}\bar{J}}\ {}^{xh}\mu, \mu') \\
&= \bigoplus_{^h\bar{J}\backslash\bar{G}/\bar{J}'} \mathrm{Hom}_{^{xh}\bar{J}\cap\bar{J}'}(\mathrm{Res}_{^{xh}\bar{J}\cap\bar{J}'}^{^{xh}\bar{J}}\ {}^{xh}\mu, \mathrm{Res}_{^{xh}\bar{J}\cap\bar{J}'}^{\bar{J}'}\ \mu').
\end{aligned}$$

Then ${}^h\mu$ and μ′ must intertwine in $\bar{G}$, and the intertwining implies conjugacy property shows that the types ${}^h\mu$ and μ′ must actually be $\bar{G}$-conjugate, and hence ${}^h\bar{\pi}$ is $\bar{G}$-conjugate to ${}^g\bar{\pi}' \simeq \bar{\pi}$. Therefore ${}^h\bar{\pi} \simeq \bar{\pi}$, and the result follows. □

We are then able to give a description of the relationship between the archetypes in the two groups $G$ and $\bar{G}$ in terms of $L$-packets.

**Proposition 3.3.** *Let π be a supercuspidal representation of G, let* s=[G,π]$_G$, *and let* ${}^G(K,\tau)$ *be the unique* s*-archetype. Let* Π *be the L-packet of irreducible components of* $\pi\downharpoonright_{\bar{G}}$. *Then the set of archetypes for the representations in* Π *is precisely the set of the* ${}^{\bar{G}}(\mathscr{K}, \bar{\tau})$, *for* $(\mathscr{K}, \bar{\tau})$ *an irreducible component of either* $\tau\downharpoonright_{\bar{K}}$ *or* ${}^\eta\tau\downharpoonright_{^\eta\bar{K}}$.

**Proof.** We show that the set of typical representations of $\bar{K}$ for some $\bar{\pi} \in \Pi$ is equal to the set of irreducible components of $\tau\downharpoonright_{\bar{K}}$; the general result then follows immediately. Let $(\bar{K}, \bar{\tau})$ be an archetype for some $\bar{\pi} \in \Pi$. Applying Lemma 2.3, $\mathrm{Ind}_{\bar{K}}^{K}\ \bar{\tau}$ contains a [G,π]$_G$-typical subrepresentation; by Frobenius reciprocity $\bar{\tau}$ is then of the required form. Conversely, the irreducible components of $\tau\downharpoonright_{\bar{K}}$ are all $K$-conjugate by Clifford theory, and so if one of them is a type for some element of Π then they all must be. ApplyingLemma 3.2, at least one of these irreducible components must be a type for some $\bar{\pi} \in \Pi$. □

**Corollary 3.4 Inertial local Langlands correspondence for $SL_2(F)$.**

*Let* $\varphi: I_F \rightarrow \mathbf{PGL}_2(\mathbb{C})$ *be a representation extending to an irreducible L-parameter* $\tilde{\varphi}: W_F \rightarrow \mathbf{PGL}_2(\mathbb{C})$. *Then there exists a finite set* $\{(K_i, \tau_i)\}$ *of smooth irreducible representations* $\tau_i$ *of maximal compact subgroups* $K_i$ *of* $\bar{G}$ *such that, for all smooth, irreducible, infinite-dimensional representations* $\pi$ *of* $\bar{G}$, *we have that* $\pi$ *contains some* $\tau_i$ *upon restriction to* $K_i$ *if and only if* $\overline{\mathrm{rec}}(\pi) \lfloor_{I_F} \simeq \varphi$. *Furthermore, this set is unique up to* $\bar{G}$*-conjugacy.*

**Proof.** Let $\Pi = \overline{\mathrm{rec}}^{-1}(\tilde{\varphi})$ be the $L$ -packet corresponding to $\tilde{\varphi}$, so that Π is the set of irreducible components upon restriction to $\bar{G}$ of some supercuspidal representation $\sigma$ of $G$. Let $\psi = \mathrm{rec}(\sigma)$, so that, by Corollary 8.2 of [15], there exists a unique smooth irreducible representation $\tau$ of $K$ such that, for all smooth, irreducible, infinite-dimensional representations $\rho$ of $G$, we have that $\rho$ contains $\tau$ upon restriction to $K$ if and only if $\mathrm{rec}(\rho) \lfloor_{I_F} \simeq \psi \lfloor_{I_F}$. Then ${}^{G}(K, \tau)$ is the unique archetype for $\sigma$, and the set $\{{}^{\bar{G}}(\mathcal{K}_i, \tau_i)\}$ of archetypes for Π is precisely that represented by the finite set of irreducible components of $\tau \lfloor_{\bar{K}}$ and ${}^{\eta}\tau \lfloor_{{}^{\eta}\bar{K}}$. Let $\mathfrak{S}$ be the set of $\bar{G}$-inertial equivalence classes of representations in Π. Then, as each of the $(K_i, \tau_i)$ is an archetype, it follows that, for all smooth, irreducible, infinite-dimensional representations $\pi$ of $\bar{G}$, we have that $\pi$ contains one of the $\tau_i$ upon restriction to $K_i$ if and only if $[\bar{G}, \pi]_{\bar{G}} \in \mathfrak{S}$, if and only if $\pi \in \Pi$, if and only if $\overline{\mathrm{rec}}(\pi) \lfloor_{I_F} \simeq \varphi$, as required. □

## ACKNOWLEDGMENTS

This paper has resulted from my first year of PhD work at the University of East Anglia. It has benefited hugely from discussions with my advisor, Shaun Stevens, who is due thanks for his constant encouragement and generosity with ideas. I am also grateful to the EPSRC for providing my research studentship, and to the anonymous referee for some helpful comments.


## REFERENCES


1. J.N. Bernstein, Le “centre” de Bernstein, in: P. Deligne (Ed.), Representations of Reductive Groups over a Local Field, Hermann, Paris, 1984, pp. 1–32.

2. C. Breuil, A. Mézard, Multiplicités modulaires et représentations de GL2(Zp) et de Gal(Qp/Qp) en l = p, Duke Math. J. 115 (2) (2002) 205–310,

3. C.J. Bushnell, G. Henniart, The Local Langlands Conjecture for GL(2), Grundlehren Math. Wiss., vol. 335, Springer-Verlag, Berlin, 2006.

4. C.J. Bushnell, P.C. Kutzko, The Admissible Dual of GL(N) via Compact Open Subgroups, Ann. of Math. Stud., vol. 129, Princeton University Press, Princeton, NJ, 1993.

5. C.J. Bushnell, P.C. Kutzko, The admissible dual of SL(N). I, Ann. Sci. Éc. Norm. Supér. (4) 26 (2) (1993) 261–280, http://www.numdam.org/item?id=ASENS_1993_4_26_2_261_0.

6. C.J. Bushnell, P.C. Kutzko, The admissible dual of SL(N). II, Proc. Lond. Math. Soc. (3) 68 (2) (1994) 317–379,

7. C.J. Bushnell, P.C. Kutzko, Smooth representations of reductive p-adic groups: structure theory via types, Proc. Lond. Math. Soc. (3) 77 (3) (1998) 582–634,

8. C.J. Bushnell, P.C. Kutzko, Semisimple types in GLn, Compos. Math. 119 (1) (1999) 53–97,

9. S.S. Gelbart, A.W. Knapp, L-indistinguishability and R groups for the special linear group, Adv. Math. 43 (2) (1982) 101–121,

10. D. Goldberg, A. Roche, Types in SLn, Proc. Lond. Math. Soc. (3) 85 (1) (2002) 119–138,

11. J.-L. Kim, Supercuspidal representations: an exhaustion theorem, J. Amer. Math. Soc. 20 (2) (2007) 273–320, http:// dx. doi. org/10.1090/S0894-0347-06-00544-3 (electronic).

12. J.-P. Labesse, R.P. Langlands, L-indistinguishability for SL(2), Canad. J. Math. 31 (4) (1979) 726–785,

13. M. Miyauchi, S. Stevens, Semisimple types for p-adic classical groups, Math. Ann. 358 (1–2) (2014) 257–288,

14. S. Nadimpalli, Typical representations of GL3(F), arXiv:1405.7286.

15. V. Paskunas, Unicity of types for supercuspidal representations of GLN , Proc. Lond. Math. Soc. (3) 91 (3) (2005) 623–654,

16. V. Sécherre, Représentations lisses de GLm(D). III. Types simples, Ann. Sci. Éc. Norm. Supér. (4) 38 (6) (2005) 951–977,

17. V. Sécherre, S. Stevens, Représentations lisses de GLm(D). IV. Représentations supercuspidales, J. Inst. Math. Jussieu 7 (3) (2008) 527–574,

18. V. Sécherre, S. Stevens, Smooth representations of GLm(D) VI: semisimple types, Int. Math. Res. Not. IMRN 13 (2012) 2994–3039.

19. S. Stevens, The supercuspidal representations of p-adic classical groups, Invent. Math. 172 (2) (2008) 289–352,

20. J.-K. Yu, Construction of tame supercuspidal representations, J. Amer. Math. Soc. 14 (3) (2001) 579–622,

2

# SPECIAL NUMBERS ON ANALYTIC FUNCTIONS

**Yilmaz Simsek**

Department of Mathematics, Faculty of Science,
University of Akdeniz, Antalya, Turkey

## ABSTRACT

The aim of this paper is to give some analytic functions which are related to the generating functions for the central factorial numbers. By using these functions and p-adic Volkenborn integral, we derive many new identities associated with the Bernoulli and Euler numbers, the central factorial numbers and the Stirling numbers. We also give some remarks and comments on these analytic functions, which are related to the generating functions for the special numbers.

## KEYWORDS

Bernoulli Numbers, Euler Numbers, The Central Factorial Numbers , Array Polynomials, Stirling Numbers of the First Kind and the Second Kind, Generating Function, Functional Equation, Analytic Functions

# 1. INTRODUCTION, DEFINITIONS AND PRELIMINARIES

Throughout this paper, we use the following standard notations:

$\square = \{1,2,3,\cdots\}$, $\square_0 = \{0,1,2,3,\cdots\} = \square \cup \{0\}$ and $\square^- = \{-1,-2,-3,\cdots\}$. Here, z denotes the set of integers, rdenotes the set of real numbers and c denotes the set of complex numbers. We assume that $\ln(z)$ denotes the principal branch of the multi-valued function $\ln(z)$ with the imaginary part $\Im(\ln(z))$ constrained by $-\pi < \Im(\ln(z)) \leq \pi$. Furthermore, $0^n = 1$ if $n = 0$, and, $0^n = 0$ if $n \in \square$ .

$$\binom{x}{v} = \frac{x(x-1)\cdots(x-v+1)}{v!}$$

(cf. [1] -[19] ).

The generating functions have been used in many branches of Mathematics and Mathematical Physics. By using p-adic Volkenborn integral, analytic functions and these generating functions, we derive many functional equations and differential equations. By using these equations, we derive many new identities and combinatorics relations related to these numbers and polynomials.

The Stirling numbers of the first and the second are used in pure and applied Mathematics. The Stirling number of the second kind, denoted by $S(n,k)$, is the number of ways to partition a set of $n$ objects into $k$ groups. These numbers occur in combinatorics and in the theory of partitions. The Stirling numbers of the second kind $S(n,v)$ are defined by means of the following generating function:

$$F_S(t,v) = \frac{\left(e^t - 1\right)^v}{v!} = \sum_{n=0}^{\infty} S(n,v)\frac{t^n}{n!},$$

(cf. [1] -[19] ). These numbers are computed by the following formula:

$$x^n = \sum_{v=0}^{n} \binom{x}{v} v! S(n,v)$$

or

$$S(n,v) = \frac{1}{v!} \sum_{j=0}^{v} \binom{v}{j} (-1)^j (v-j)^n .$$

From the above equation, one has

$$S(n,k) = S(n-1,k-1) + kS(n-1,k),$$

with

$$S(n,0) = 0 (n \in \square); S(n,n) = 1 (n \in \square); S(n,1) = 1 (n \in \square)$$

and $S(n,k)=0$ ($n<k$ or $k<0$) (cf. [1] -[19] ).

The Stirling numbers of the first kind $s(n,k)$ are defined by means of the following generating function:

$$\frac{(\log(1+x))^v}{v!} = \sum_{n=0}^{\infty} s(n,v) \frac{t^n}{n!},$$

(cf. [19] ). These numbers are computed by the following formula:

$$x(x-1)\cdots(x-n+1) = \sum_{v=0}^{n} s(n,v) x^v .$$

From the above equation, one has

$$s(n,k) = s(n-1,k-1) - (n-1)s(n-1,k),$$

with

$$s(n,0)=0(n\in\mathbb{N});s(n,n)=1(n\in\mathbb{N}_0);s(n,1)=(-1)^{n-1}(n-1)!(n\in\mathbb{N})$$

and $s(n,k)=0$ $(n<k$ or $k<0)$ (cf. [19] ).

The array polynomials $S_v^n(x)$ are defined by means of the following generating function:

$$\frac{\left(e^t-1\right)^v}{v!}e^{tx}=\sum_{n=0}^{\infty}S_v^n(x)\frac{t^n}{n!},$$

(cf. [1] [13] ).

From the above generating function, one has

$$S_v^n(x)=\frac{1}{v!}\sum_{j=0}^{v}(-1)^{v-j}\binom{v}{j}(x+j)^n$$

with

$$S_0^0(x)=S_n^n(x)=1,S_0^n(x)=x^n$$

and for $v>n$, $S_v^n(x)=0$ (cf. [1] [13] ).

## 2. IDENTITIES AND RELATIONS ON THE CENTRAL FACTORIAL NUMBERS $T(n,k)$

In this section, we study on generating functions for the central factorial numbers $T(n,k)$ on analytic functions. By using these functions, we derive are some functional equations. Applying these functions and equations, we give some identities and relations related to the central factorial numbers $T(n,k)$, which are defined by means of the following generating function (cf. [3] [19] ):

$$F_T(t,k)=\frac{1}{(2k)!}\left(e^t+e^{-t}-2\right)^k=\sum_{n=k}^{\infty}T(n,k)\frac{t^{2n}}{(2n)!}. \tag{2.1}$$

Srivastava and Liu [19] gave many properties and applications of the central factorial numbers.

In the work of Srivastava and Liu [19] , one has the following relations:

$$x^n=\sum_{k=0}^{n}T(n,k)x(x-1)(x-2^2)(x-3^2)\cdots(x-(k-1)^2).$$

Combining the above equation with (1), we also have

$$T(n,k)=T(n-1,k-1)+k^2T(n-1,k),$$

where $n\geq 1$, $k\geq 1$, $(n,k)\neq(1,1)$. For $n$, $k\in\square$, $T(0,k)=T(n,0)=0$ and $T(n,1)=1$ (cf. [19] ).

In [2] , Cigler gave the following formula for the numbers $T(n,k)$:

$$T(n,k)=\frac{1}{(2k)!}\sum_{j=0}^{2k}(-1)^j\binom{2k}{j}(j-k)^{2n}.$$

Some properties of the function $F_T(t,k)$ are given as follows:

$$F_T(0,k)=1.$$

From this property, one has $T(0,0)=1$. The function $F_T(t,k)$ is related to the hyperbolic functions and generating functions for the special numbers. Thus we get the following functional equation:

$$F_T(t,k)=\frac{2^k}{(2k)!}(\cosh(t)-1)^k$$

The function $F_T(t,k)$ is related to the generating function for the Stirling numbers of the second kind, $S(n,k)$. That is

$$F_T(t,k) = F_S(t,2k)e^{-kt}, \tag{2.2}$$

where

$$F_S(t,2k) = \frac{1}{(2k)!}\left(e^t - 1\right)^{2k} = \sum_{n=0}^{\infty} S(n,2k)\frac{t^n}{n!}.$$

and

$$F_T(t,k) = \frac{k!}{(2k)!}\sum_{j=0}^{k} F_S(t,j)F_S(-t,k-j). \tag{2.3}$$

We are going to differentiate (1) with respect to $t$ to derive a recurrence relation or identities for the central factorial numbers.

$$\frac{d}{dt}F_T(t,k) = \frac{k(2k-2)!}{(2k)!}\left(e^t - e^{-t}\right)F_T(t,k-1) \tag{2.4}$$

or

$$\frac{d}{dt}F_T(t,k) = \frac{2k(2k-2)!}{(2k)!}F_T(t,k-1)\sinh(t). \tag{2.5}$$

Lemma 1 ([9] , Lemma 11, Equation (7)]

$$\sum_{n=0}^{\infty}\sum_{k=0}^{\infty} A(n,k) = \sum_{n=0}^{\infty}\sum_{k=0}^{\left[\frac{n}{2}\right]} A(n,n-2k),$$

where $[x]$ denotes the greatest integer function.

Remark 1 Lemma 1 was generalized by Srivastava et al. [16] . By using these generalization, one may generalize Theorem 1.

**Theorem 1**

$$S(n,2k)=n!\sum_{j=0}^{\left[\frac{n}{2}\right]}\frac{k^{n-2j}}{(2j)!(n-2j)!}T(j,k).$$

Proof. By combining (2) with (1), we get

$$\sum_{n=0}^{\infty}S(n,2k)\frac{t^n}{n!}=\sum_{n=0}^{\infty}\frac{(kt)^n}{n!}\sum_{n=0}^{\infty}T(n,k)\frac{t^{2n}}{(2n)!}.$$

By using Lemma 1, we obtain

$$\sum_{n=0}^{\infty}S(n,2k)\frac{t^n}{n!}=\sum_{n=0}^{\infty}\left(\sum_{j=0}^{\left[\frac{n}{2}\right]}\frac{k^{n-2j}}{(2j)!(n-2j)!}T(j,k)\right)t^n.$$

Comparing the coefficients of $t^n$ on both sides of the above equation, we arrive at the desired result. □

**Theorem 2**

$$\sum_{j=0}^{\left[\frac{n}{2}\right]}\frac{k^{n-2j}}{(2j)!(n-2j)!}T(j,k)=\frac{k!}{n!(2k)!}\sum_{j=0}^{k}\sum_{l=0}^{n}(-1)^{k-j}\binom{n}{l}S_j^l(j)S(n-l,k-j),$$

where $S_j^l(j)$ denotes the array polynomials.

Proof. By using (3), we derive the following functional equation:

$$F_T(t,k)e^{tx}=\frac{k!}{(2k)!}\sum_{j=0}^{k}(-1)^{k-j}F_A(t,j)F_S(t,k-j), \tag{2.6}$$

where $F_A(t,j)$ denotes generating function for the array polynomials $S_j^n(x)$. By combining (2.6) with (2.1), we obtain

$$\frac{k!}{(2k)!}\sum_{j=0}^{k}(-1)^{k-j}\sum_{n=0}^{\infty}S_j^n(j)\frac{t^n}{n!}\sum_{n=0}^{\infty}S(n,k-j)\frac{t^n}{n!}=\sum_{n=0}^{\infty}\frac{(kt)^n}{n!}\sum_{n=0}^{\infty}T(n,k)\frac{t^{2n}}{(2n)!}.$$

By using Lemma 1, we obtain

$$\sum_{n=0}^{\infty}\left(\sum_{j=0}^{\left[\frac{n}{2}\right]}\frac{k^{n-2j}}{(2j)!(n-2j)!}T(j,k)\right)t^n=\sum_{n=0}^{\infty}\left(\frac{k!}{n!(2k)!}\sum_{j=0}^{k}\sum_{l=0}^{n}(-1)^{k-j}\binom{n}{l}S_j^l(j)S(n-l,k-j)\right)\frac{t^n}{n!}.$$

Comparing the coefficients of $t^n$ on both sides of the above equation, we arrive at the desired result.

Remark 2 By using functional equation in (2.3), (2.4) and (2.5), one may also obtain an interesting identities and relationships between the Stirling numbers of the second kind and the central factorial numbers.

# 3. IDENTITIES RELATED TO THE P-ADIC VOLKENBORN INTEGRAL AND THE CENTRAL FACTORIAL NUMBERS

In this section, we give applications of the p-adic integral on $\mathbb{Z}_p$, set of p-adic integers. By using the Witt's formula for the Bernoulli and Euler numbers related to the p-adic Volkenborn integral and the Mahler coefficients, we derive many interesting and novel identities associated with the central factorial numbers, the family of numbers $C_n^{(x)}$, the Stirling numbers of the first and the second kind and also combinatorics identities.

We use notations of the work of Kim [4] for the $p$-adic $q$-Volkenborn integral. Let $p$ be a fixed prime. It is known that

$$\mu_q\left(x+p^N\mathbb{Z}_p\right)=\frac{q^x}{\left[p^N\right]_q}$$

is a distribution on $\mathbb{Z}_p$ for $q\in\mathbb{Z}_p$ with $|1-q|_p<1$, cf. [4] . Let $UD(\mathbb{Z}_p)$ be the set of uniformly differentiable functions on $\mathbb{Z}_p$. The $p$-adic $q$-integral of the function $f\in UD(\mathbb{Z}_p)$ is defined by Kim [4] as follows:

$$\int_{\mathbb{Z}_p} f(x)\,\mathrm{d}\mu_q(x)=\lim_{N\to\infty}\frac{1}{\left[p^N\right]_q}\sum_{x=0}^{p^N-1}f(x)q^x,$$

where

$$[x]=\frac{1-q^x}{1-q}.$$

The bosonic $p$-adic integral ($p$-adic Volkenborn integral) was considered from a physical point of view to the bosonic limit $q\to 1$, as follows ([4] ):

$$\int_{\mathbb{Z}_p} f(x)\,\mathrm{d}\mu_1(x)=\lim_{N\to\infty}\frac{1}{p^N}\sum_{x=0}^{p^N-1}f(x), \tag{3.1}$$

where

$$\mu_1\left(x+p^N\mathbb{Z}_p\right)=\frac{1}{p^N}.$$

The $p$-adic $q$-integral is used in many branch of mathematics, mathematical physics and other areas (cf. [4] [11] [14] [18] ).

By using (7), we have the Witt's formula for the Bernoulli numbers $B_n$ as follows:

$$\int_{\mathbb{Z}_p} x^n d\mu_1(x) = B_n \tag{3.2}$$

(cf. [4] [5] [11] [14] [18] ).

We consider the fermionic integral in contrast to the bosonic integral, which is called the fermionic $p$-adic Volkenborn integral on $\mathbb{Z}_p$ cf. [5] . That is

$$\int_{\mathbb{Z}_p} f(x) d\mu_{-1}(x) = \lim_{N\to\infty} \sum_{x=0}^{p^N-1} (-1)^x f(x) \tag{3.3}$$

where

$$\mu_1\left(x + p^N \mathbb{Z}_p\right) = \frac{(-1)^x}{p^N}$$

(cf. [5] ). By using (9), we have the Witt's formula for the Euler numbers $E_n$ as follows:

$$\int_{\mathbb{Z}_p} x^n d\mu_{-1}(x) = E_n, \tag{3.4}$$

(cf. [5] [6] [11] [18] ).

**Theorem 3**

$$\int_{\mathbb{Z}_p} \binom{x}{j} d\mu_1(x) = \frac{(-1)^j}{j+1}. \tag{3.5}$$

Proof of Theorem 3 was given by Schikhof [14].

**Theorem 4**

$$\int_{\mathbb{Z}_p} \binom{x}{j} d\mu_{-1}(x) = \frac{(-1)^j}{2^j}.$$

Proof of Theorem 4 was given by Kim et al. ([7] [8] ).

The family of numbers $C_n^{(x)}$ was defined by Srivastava and Liu [19] . These numbers are related to the Stirling numbers of the first and the second kind, which are given by the following theorem.

Theorem 5 ([19] , Theorem 1]) Let $n \geq k$ $(n,k \in \mathbb{N})$ and

$$w(n,k) = (-1)^k \sum_{j=k}^{n} 2^{n-j} S(n,j) s(j,k).$$

Then

$$C_n^{(x)} = \sum_{k=1}^{n} w(n,k) x^k. \tag{3.6}$$

In [19] , Equation-(2.3)], Srivastava and Liu also gave the following relation:

$$C_n^{(x)} = \sum_{j=0}^{n} (-1)^j j! 2^{n-j} \binom{x+j-1}{j} S(n,j). \tag{3.7}$$

By using the above formulae, we derive the following identities related to the numbers $C_n^{(x)}$, the Stirling numbers, the Bernoulli numbers and the Euler numbers.

According to the work of Kim at al. ([7] [8] ), we get

$$\int_{\mathbb{Z}_p} \binom{-x}{j} d\mu_{-1}(x) = \int_{\mathbb{Z}_p} \binom{x+j-1}{j} d\mu_{-1}(x).$$

Combining Theorem 4 with the above equation, we get

$$\int_{\mathbb{Z}_p} \binom{x+n-1}{n} d\mu_{-1}(x) = \sum_{m=0}^{n} (-1)^m \binom{n-1}{n-m} 2^{-m}. \tag{3.8}$$

Integrating both sides of (3.7) with respect to the fermionic integral on $\mathbb{Z}_p$, and than using (3.8), we obtain the following Lemma:

**Lemma 2**

$$\int_{\mathbb{Z}_p} C_n^{(x)} d\mu_{-1}(x) = \sum_{j=0}^{n} (-1)^j j! S(n,j) \sum_{m=0}^{j} (-1)^m \binom{j-1}{j-m} 2^{n-m-j}.$$

Integrating both sides of (3.6) with respect to the fermionic integral on $\mathbb{Z}_p$, and than using (3.4), we obtain the following Lemma:

**Lemma 3**

$$\int_{\mathbb{Z}_p} C_n^{(x)} d\mu_{-1}(x) = \sum_{k=0}^{n} E_k w(n,k),$$

where $E_k$ denotes the Euler numbers.

By using Lemma 2 and Lemma 3, we arrive at the following theorem:

**Theorem 6**

$$\sum_{k=0}^{n} E_k w(n,k) = \sum_{j=0}^{n} (-1)^j j! S(n,j) \sum_{m=0}^{j} (-1)^m \binom{j-1}{j-m} 2^{n-m-j}.$$

Similarly, according to the work of Kim et al. ([7] [8] ), we get

$$(-1)^n \int_{\mathbb{Z}_p} \binom{-x}{j} d\mu_1(x) = \int_{\mathbb{Z}_p} \binom{x+n-1}{n} d\mu_1(x).$$

Combining Theorem 3 with the above equation, we get

$$\int_{\mathbb{Z}_p} \binom{x+j-1}{j} d\mu_1(x) = \sum_{m=0}^{j} (-1)^m \binom{j-1}{j-m} \frac{1}{m+1}. \tag{3.9}$$

Integrating both sides of (3.7) with respect to the bosonic $p$-adic integral on $\mathbb{Z}_p$, and than using (3.9), we obtain the following lemma:

**Lemma 4**

$$\int_{\mathbb{Z}_p} C_n^{(x)} d\mu_1(x) = \sum_{j=0}^{n} (-1)^j j! S(n,j) \sum_{m=0}^{j} (-1)^m \binom{j-1}{j-m} \frac{2^{n-j}}{m+1}.$$

Integrating both sides of (3.6) with respect to the fermionic integral on $\mathbb{Z}_p$, and than using (3.2), we obtain the following lemma:

**Lemma 5**

$$\int_{\mathbb{Z}_p} C_n^{(x)} d\mu_1(x) = \sum_{k=0}^{n} B_k w(n,k),$$

where $B_k$ denotes the Bernoulli numbers.

By using Lemma 4 and Lemma 5, we arrive at the following theorem:

**Theorem 7**

$$\sum_{k=0}^{n} B_k w(n,k) = \sum_{j=0}^{n} (-1)^j j! S(n,j) \sum_{m=0}^{j} (-1)^m \binom{j-1}{j-m} \frac{2^{n-j}}{m+1}.$$

## ACKNOWLEDGEMENTS

The paper was supported by the Scientific Research Project Administration of Akdeniz University.

# REFERENCES

1. Chang, C.-H. and Ha, C.-W. (2006) A Multiplication Theorem for the Lerch Zeta Function and Explicit Representations of the Bernoulli and Euler polynomials. Journal of Mathematical Analysis and Applications, 315, 758-767.

2. Cigler, J. Fibonacci Polynomials and Central Factorial Numbers. Preprint.http://homepage.univie.ac.at/johann.cigler/preprints/central-factorial.pdf

3. Comtet, L. (1974) Advanced Combinatorics: The Art of Finite and Infinite Expansions. Reidel, Dordrecht and Boston, (Translated from the French by J. W. Nienhuys).

4. Kim, T. (2002) q-Volkenborn integration. Russian Journal of Mathematical Physics, 19, 288-299.

5. Kim, T. (2006) q-Euler Numbers and Polynomials Associated with p-Adic q-Integral and Basic q-zeta Function. Trends in International Mathematics and Science Study, 9, 7-12.

6. Jang, L.C. and Kim, T. (2008) A New Approach to q-Euler Numbers and Polynomials. Journal of Concrete and Applicable Mathematics, 6, 159-168.

7. Kim, D.S. and Kim, T. (2013) Daehee Numbers and Polynomials. Applied Mathematical Sciences, 7, 5969-5976.

8. Kim, D.S., Kim, T. and Seo, J. (2013) A Note on Changhee Numbers and Polynomials. Advanced Studies in Theoretical Physics, 7, 993-1003.

9. Rainville, E.D. (1960) Special Functions. The Macmillan Company, New York.

10. Simsek, Y. (2010) On q-Deformed Stirling numbers. International Journal of Computer Mathematics, 15, 70-80.

11. Simsek, Y. (2010) Complete sum of products of (h,q)-Extension of Euler Polynomials and Numbers. Journal of Difference Equations and Applications, 16, 1331-1348.

12. Simsek, Y. (2013) Identities Associated with Generalized Stirling Type Numbers and Eulerian Type Polynomials. Mathematical and Computational Applications, 18, 251-263.

13. Simsek, Y. (2013) Generating Functions for Generalized Stirling type Numbers, Array Type Polynomials, Eulerian Type Polynomials and Their Applications. Fixed Point Theory and Applications, 87, 343-1355.

14. Schikhof, W.H. (1984) Ultrametric Calculus: An Introduction to p-Adic Analysis. Cambridge Studies in Advanced Mathematics 4, Cambridge University Press, Cambridge.

15. Srivastava, H.M. (2011) Some Generalizations and Basic (or q-) Extensions of the Bernoulli, Euler and Genocchi Polynomials. Applied Mathematics & Information Sciences, 5, 390-444.

16. Srivastava, H.M., Ozarslan, M.A. and Kaanoğlu, C. (2010) Some Families of Generating Functions for a Certain Class of Three-Variable Polynomials. Integral Transforms and Special Functions, 21, 885-896.

17. Srivastava, H.M. and Choi, J. (2012) Zeta and q-Zeta Functions and Associated Series and Integrals. Elsevier Science Publishers, Amsterdam, London and New York.

18. Srivastava, H.M., Kim, T. and Simsek, Y. (2005) q-Bernoulli Numbers and Polynomials Associated with Multiple q-Zeta Functions and Basic L-Series. Russian Journal of Mathematical Physics, 12, 241-268.

19. Srivastava, H.M. and Liu, G.-D. (2009) Some Identities and Congruences Involving a Certain Family of Numbers. Russian Journal of Mathematical Physics, 16, 536-542.

3

# UNIFIED REPRESENTATION OF THE FAMILY OF $L$-FUNCTIONS

**Hacer Ozden1* and Yilmaz Simsek2**

1Department of Mathematics, Faculty of Art and Science, University of Uludag, Bursa, Turkey.

2Department of Mathematics, Faculty of Science, Akdeniz University, Campus, Antalya, 07058, Turkey.

## ABSTRACT

The aim of this paper is to unify the family of $L$-functions. By using the generating functions of the Bernoulli, Euler and Genocchi polynomials, we construct unification of the $L$-functions. We also derive new identities related to these functions. We also investigate fundamental properties of these functions.

## KEYWORDS

Bernoulli numbers, Bernoulli polynomials, Euler numbers, Euler polynomials, Genocchi numbers, Genocchi polynomials, Dirichlet, $L$-functions, Hurwitz zeta function, Riemann zeta function

# 1 INTRODUCTION

The theory of the family of $L$-functions has become a very important part in the analytic number theory. In this paper, using a new type generating function of the family of special numbers and polynomials, we construct unification of the $L$-functions.

Throughout this presentation, we use the following standard notions N={1,2,...}, N0={0,1,2,...}=N∪{0}, Z+={1,2,3,...}$\mathbb{Z}^+ = \{1, 2, 3, \ldots\}$, Z−={−1,−2,...}. Also, as usual $\mathbb{Z}$ denotes the set of integers, $\mathbb{R}$ denotes the set of real number and$\mathbb{C}$ denotes the set of complex numbers. We assume that ln(z)$\ln(z)$ denotes the principal branch of the multi-valued function ln(z)$\ln(z)$with the imaginary part $\Im(\ln(z))$. constrained by $-\pi < \Im(\ln(z)) \leq \pi$..

Recently, the first author [1] introduced and investigated the following generating functions which give a unification of the Bernoulli polynomials, Euler polynomials and Genocchi polynomials:

$$g_{a,b}(x;t,k,\beta) := \frac{2^{1-k}t^k e^{tx}}{\beta^b e^t - a^b} = \sum_{n=0}^{\infty} \mathcal{Y}_{n,\beta}(x;k,a,b)\frac{t^n}{n!}, \tag{1}$$

where (|t|<2π when $\beta = a$; $|t| < |b \log(\frac{\beta}{a})|$ when $\beta \neq a$; $k \in \mathbb{N}_0$; $\beta \in \mathbb{C}$ ($|\beta| < 1$); $a, b \in \mathbb{C}\backslash\{0\}$).

For the special values of $a$, $b$, $k$, $b$ and $\beta$, the polynomials Yn,β(x;k,a,b) provide us with a generalization and unification of the classical Bernoulli polynomials, Euler polynomials and Genocchi polynomials and also of the Apostol-type (Apostol-Bernoulli, Apostol-Euler, Apostol-Genocchi) polynomials.

**Remark 1.1** If we set k=a=b=1 in (1), we get a special case of the generalized Bernoulli polynomials $\mathcal{Y}_{n,\beta}(x,k,1,1)$, that is, the so-called Apostol-Bernoulli polynomials $Bn$(x,$\beta$) generated by

$$\frac{t}{\beta e^t - 1} e^{xt} = \sum_{n=0}^{\infty} \mathcal{B}_n(x,\beta) \frac{t^n}{n!}$$

(*cf.* [1, 2, 3, 4, 5, 6, 7, 8, 9, 10, 11, 12, 13, 14, 15, 16, 17, 18, 19, 20, 21, 22, 23, 24, 25, 26, 27, 28]).

**Remark 1.2** By substituting k+1=−a=b=1$k+1=-a=b=1$ in (1), we are led to Apostol-Euler polynomials En(x,β)$\mathcal{E}_n(x,\beta)$ which are defined by means of the following generating function:

$$\frac{2}{\beta e^t + 1} e^{xt} = \sum_{n=0}^{\infty} \mathcal{E}_n(x,\beta)$$

(*cf.* [1, 2, 3, 4, 5, 6, 7, 8, 9, 10, 11, 12, 13, 14, 15, 16, 17, 18, 19, 20, 21, 22, 23, 24, 25, 26, 27, 28]).

**Remark 1.3** Setting k=−a=b=1$k=-a=b=1$ into (1), we get the Apostol-Genocchi polynomials Gn(x,β)$\mathcal{G}_n(x,\beta)$ which are defined by means of the following generating function:

$$\frac{2t}{\beta e^t + 1} e^{xt} = \sum_{n=0}^{\infty} \mathcal{G}_n(x,\beta) \frac{t^n}{n!}$$

(*cf.* [1, 2, 3, 4, 5, 6, 7, 8, 9, 10, 11, 12, 13, 14, 15, 16, 17, 18, 19, 20, 21, 22, 23, 24, 25, 26, 27, 28]).

In terms of a Dirichlet character $\chi$ of conductor f∈N$f \in \mathbb{N}$, Ozden *et al.* [16] extended and investigated the generating functions of the generalized Bernoulli, Euler and Genocchi numbers and the generalized Bernoulli, Euler and Genocchi polynomials with parameters $a, b, \beta$ and $k$. Such $\chi$-extended polynomials and $\chi$-extended numbers are useful in many areas of mathematics and mathematical physics.

**Definition 1.4** (Ozden *et al.* [[16], p.2783])

Let $\chi$ be a Dirichlet character of conductor f∈N$f \in \mathbb{N}$. Then the aforementioned $\chi$-extended generalized Bernoulli-Euler-Genocchi numbers $\mathcal{Y}_{n,\chi,\beta}(k,a,b)$ and the aforementioned $\chi$-extended generalized Bernoulli-Euler-Genocchi polynomials $\mathcal{Y}_{n,\chi,\beta}(k,a,b)$. are given by the following generating functions:

$$F_{\chi,\beta}(t;k,a,b) = 2^{1-k}t^k \sum_{j=1}^{f} \frac{\chi(j)(\frac{\beta}{a})^{bj}e^{jt}}{\beta^{bf}e^{ft} - a^{bf}} = \sum_{n=0}^{\infty} \mathcal{Y}_{n,\chi,\beta}(k,a,b)\frac{t^n}{n!}, \tag{2}$$

where $(|t| < 2\pi \text{ when } \beta = a;\ |t| < |b\log(\frac{\beta}{a})| \text{ when } \beta \neq a;\ k \in \mathbb{N}_0;\ \beta \in \mathbb{C}\ (|\beta| < 1);\ a,b \in \mathbb{C}\backslash\{0\})$ and

$$\mathfrak{H}_{\chi,\beta}(x,t;k,a,b) = F_{\chi,\beta}(t,k;a,b)e^{tx} = \sum_{n=0}^{\infty} \mathcal{Y}_{n,\chi,\beta}(x;k,a,b)\frac{t^n}{n!} \tag{3}$$

$(|t| < 2\pi \text{ when } \beta = a;\ |t| < |b\log(\frac{\beta}{a})| \text{ when } \beta \neq a;\ k \in \mathbb{N}_0;\ \beta \in \mathbb{C}\ (|\beta| < 1);\ a,b \in \mathbb{C}\backslash\{0\})$.

**Remark 1.5** Substituting k=a=b=β=1 into (2), we are led immediately to the generating function of the generalized Bernoulli numbers which are defined by means of the following generating function:

$$\sum_{j=1}^{f} \frac{\chi(j)te^{jt}}{e^{ft} - 1} = \sum_{n=0}^{\infty} B_{n,\chi}\frac{t^n}{n!} \tag{4}$$

(*cf.* [1, 2, 3, 4, 5, 6, 7, 8, 9, 10, 11, 12, 13, 14, 15, 16, 17, 18, 19, 20, 21, 22, 23, 24, 25, 26]).

# 2 UNIFICATION OF THE *L*-FUNCTIONS

Our aim in this section is to apply the Mellin transformation to the generating function (3) of the polynomials $\mathcal{Y}_{n,\chi,\beta}(x;k,a,b)$ in order to construct a unification of the various members of the family of the *L*-functions and to thereby interpolate $\mathcal{Y}_{n,\chi,\beta}(x;k,a,b)$ for negative integer values of *n*.

Throughout this section, we assume that β∈C with |β|<1 and s∈C.

By substituting (1) into (2), we obtain the following functional equation:

$$F_{\chi,\beta}(t;k,a,b)=\frac{1}{f^k}\sum_{j=1}^{f}\chi(j)\left(\frac{\beta}{a}\right)^{bj}g_{a^f,b}\left(\frac{j}{f},tf;k,\beta^f\right). \tag{5}$$

By using this functional equation, we arrive at the following theorem.

**Theorem 2.1** *Let $\chi$ be a Dirichlet character of conductor $f$. Then we have*

$$\mathcal{Y}_{n,\chi,\beta}(k,a,b)=f^{n-k}\sum_{j=1}^{f}\chi(j)\left(\frac{\beta}{a}\right)^{bj}\mathcal{Y}_{n,\beta^f}\left(\frac{j}{f};k,a^f,b\right). \tag{6}$$

By using (5), we modify (3) as follows:

$$\mathfrak{H}_{\chi,\beta}(x,t;k,a,b)=\frac{1}{f^k}\sum_{j=1}^{f}\chi(j)\left(\frac{\beta}{a}\right)^{bj}g_{a^f,b}\left(\frac{j+x}{f},tf;k,\beta^f\right). \tag{7}$$

By using (7), we derive the following result.

**Corollary 2.2** *Let $\chi$ be a Dirichlet character of conductor* f∈N. *Then we have*

$$\mathcal{Y}_{n,\chi,\beta}(x;k,a,b) = f^{n-k} \sum_{j=1}^{f} \chi(j) \left(\frac{\beta}{a}\right)^{bj} \mathcal{Y}_{n,\beta^f}\left(\frac{j+x}{f};k,a^f,b\right). \tag{8}$$

By applying the Mellin transformation to the generating function (1), Ozden *et al.* [[16], p.2784 Equation (4.1)] gave an integral representation of the unified zeta function ζβ(s,x;k,a,b):

$$\zeta_\beta(s,x;k,a,b) = \frac{1}{\Gamma(s)} \int_0^\infty t^{s-k-1} g_{a,b}(x;-t;k,\beta)\,dt \quad (\min\{\Re(s),\Re(x)\} > 0), \tag{9}$$

where the additional constraint R(x)>0 is required for the convergence of the infinite integral, which is given in (9), at its upper terminal. By making use of the above integral representation, Ozden *et al.* [[16], p.2784 Equation (4.1)] defined the unified zeta function ζβ(s,x;k,a,b) as follows:

$$\zeta_\beta(s,x;k,a,b) = \left(-\frac{1}{2}\right)^{k-1} \sum_{m=0}^{\infty} \frac{\beta^{bm}}{a^{b(m+1)}(m+x)^s} \quad (\beta \in \mathbb{C}(|\beta|<1); s \in \mathbb{C}(\Re(s)>1)). \tag{10}$$

By applying the Mellin transformation to the generating function (7), we have the following integral representation of the unified two-variable *L*-functions Lχ,β(s,x;k,a,b):

$$L_{\chi,\beta}(s,x;k,a,b) = \sum_{j=1}^{f} \frac{\chi(j)(\frac{\beta}{a})^{bj}}{f^k\Gamma(s)} \int_0^\infty t^{s-k-1} g_{a^f,b}\left(\frac{j+x}{f},-tf;k,\beta^f\right) dt$$
$$(\min\{\Re(s),\Re(x)\} > 0) \tag{11}$$

in terms of the generating function Hχ,β(x,t;k;a,b) defined in (7). By substituting (9) into ( 11), we obtain

$$L_{\chi,\beta}(s,x;k,a,b) = \frac{1}{f^{k+s}} \sum_{j=1}^{f} \chi(j) \left(\frac{\beta}{a}\right)^{bj} \zeta_{\beta^f}\left(s,\frac{j+x}{f};k,a^f,b\right) \tag{12}$$

where (β∈C (|β|<1); s∈C (R(s)>1)).

Consequently, by making use of (10) and (12), we are ready to define a two-variable unification of the Dirichlet-type *L*-functions $L_{\chi,\beta}(s,x;k,a,b)$ as follows.

**Definition 2.3** Let $\chi$ be a Dirichlet character of conductor $f$?$N$. For s,β∈C (|β|<1), we define a two-variable unified *L*-function $L_{\chi,\beta}(s,x;k,a,b)$ by

$$L_{\chi,\beta}(s,x;k,a,b) = f^{-k}\left(-\frac{1}{2}\right)^{k-1}\sum_{m=0}^{\infty}\frac{\beta^{bm}\chi(m)}{a^{b(m+f)}(m+x)^s} \quad (\beta\in\mathbb{C}\ (|\beta|<1);\Re(s)>1). \tag{13}$$

**Remark 2.4** If we substitute x=1$x=1$ into (13), we get the unified *L*-function

$$L_{\chi,\beta}(s;k,a,b) := L_{\chi,\beta}(s,1;k,a,b)$$

by

$$L_{\chi,\beta}(s;k,a,b) = f^{-k}\left(-\frac{1}{2}\right)^{k-1}\sum_{m=1}^{\infty}\frac{\beta^{bm}\chi(m)}{a^{b(m+f)}m^s},$$

where $(\Re(s)>1,\ \beta\in\mathbb{C}\ (|\beta|<1))$.

**Remark 2.5** Upon substituting k=a=b=1 and $\beta=\frac{\xi}{u}$ into (13), we arrive at the interpolation function for twisted generalized Eulerian numbers and polynomials, which is given as follows:

$$l_1\left(\frac{u}{\xi},s,\chi\right) = L_{\chi,\frac{\xi}{u}}(s,x;1,1,1),$$

where, for a positive integer *r*, $\xi$ is the *r* th root of 1.

$$l_1\left(\frac{u}{\xi},s;\chi\right) = \sum_{m=0}^{\infty}\left(\frac{\xi}{u}\right)^m\frac{\chi(m)}{(m+x)^s}$$

(*cf.* [18]).

**Remark 2.6** Substituting x=1$x=1$ into (13), we get a unification of the $L$-functions

$$L_{\chi,\beta}(s,1;k,a,b) = L_{\chi,\beta}(s;k,a,b).$$

Substituting $\chi\equiv 1$ into (13), we get a unification $\zeta^{\beta}(s,x;k,a,b)$ of the Hurwitz-type zeta function which is given in (10). We also note that both the Hurwitz (or generalized) zeta function

$$\zeta(s,x) = \zeta_1(s,x;1,1,1) = \sum_{n=0}^{\infty} \frac{1}{(n+x)^s}$$

(*cf.* [27, 28]) and the Riemann zeta function

$$\zeta(s) = \zeta_1(s,1;1,1,1) = \sum_{n=1}^{\infty} \frac{1}{n^s}$$

are obvious special cases of the unified zeta function $\zeta^{\beta}(s,x;k,a,b)$ (*cf.* [16, 27, 28]). *The relationship between the unified zeta function and the Hurwitz-Lerch zeta function* Φ(z,s,a)$\Phi(z,s,a)$ *was given by Ozden et al.* [16]*:*

$$\zeta_{\beta}(s,x;k,a,b) := \left(-\frac{1}{2}\right)^{k-1} a^{-b}\Phi\left(\frac{\beta^b}{a^b}, s, x\right), \tag{14}$$

where *the Hurwitz-Lerch zeta function is defined by*

$$\Phi(z,s,x) = \sum_{n=0}^{\infty} \frac{z^n}{(n+x)^s},$$

which converges for $(x \in \mathbb{C}\setminus\mathbb{Z}_0^-, s \in \mathbb{C} \text{ when } |z|<1; \Re(s)>1 \text{ when } |z|=1)$, where as usual

$$\mathbb{Z}_0^- = \mathbb{Z}^- \cup \{0\}$$

(*cf.* [27, 28]).

A relationship between the functions $L_{\chi,\beta}(s,x;k,a,b)$ and $\zeta_\beta(s,x;k,a,b)$ is provided by the next theorem.

**Theorem 2.7** *Let* s∈C. *Let* $\chi$ *be a Dirichlet character of conductor* f∈N. *Then we have*

$$L_{\chi,\beta}(s,x;k,a,b) = f^{-s-k}\sum_{j=1}^{f}\left(\frac{\beta}{a}\right)^{jb}\chi(j)\zeta_{\beta^f}\left(s,\frac{j+x}{f};k,a^f,b\right). \tag{15}$$

*Proof* Substituting *m=nf+j, j=1,2,...,f, n=0,...,*∞ into (13), we obtain

$$L_{\chi,\beta}(s,x;k,a,b) = \left(-\frac{1}{2}\right)^{k-1} f^{-s-k}\sum_{j=1}^{f}\left(\frac{\beta}{a}\right)^{jb}\chi(j)\sum_{n=0}^{\infty}\frac{\beta^{bnf}}{a^{bnf}(n+\frac{j+x}{f})^s}.$$

After some algebraic manipulations, we arrive at the desired result.
□

**Remark 2.8** Substituting a=b=k=1 into (13), we have

$$L_{\chi,\beta}(s,x;1,1,1) = \sum_{m=0}^{\infty}\frac{\beta^m\chi(m)}{(m+x)^s} \quad (\Re(s) > 1, \beta \in \mathbb{C}(|\beta| < 1))$$

which interpolates the Apostol-Bernoulli polynomials attached to the Dirichlet character, which are given by means of the following generating functions:

$$\sum_{j=1}^{f}\frac{\chi(j)t\beta^j e^{t(j+x)}}{\beta^f e^{tf}-1} = \sum_{n=0}^{\infty}\mathcal{B}_{n,\chi}(x,\beta)\frac{t^n}{n!}.$$

Let *f* be an odd integer. If we set a=−1 and k=0 into (13), then we have

$$L_{\chi,\beta}(s,x;1,-1,1)=2\sum_{m=1}^{\infty}(-1)^m\frac{\chi(m)\beta^m}{(m+x)^s}\quad(\Re(s)>1,\beta\in\mathbb{C}(|\beta|<1)),$$

which interpolate the Apostol-Euler polynomials attached to the Dirichlet character, which are defined by the following generating functions:

$$\sum_{j=1}^{f}\frac{2\chi(j)\beta^j e^{t(j+x)}}{\beta^f e^{tf}+1}=\sum_{n=0}^{\infty}\mathcal{E}_{n,\chi}(x,\beta)\frac{t^n}{n!}$$

(*cf.* [1, 2, 3, 4, 5, 6, 7, 8, 9, 10, 11, 12, 13, 14, 15, 16, 17, 18, 19, 20, 21, 22, 23, 24, 25, 26, 27, 28, 29]).

By using (15) and (14), we arrive at the following result.

**Corollary 2.9** *Let* s∈C. *Let* $\chi$ *be a Dirichlet character of conductor* f∈N. *Then we have*

$$L_{\chi,\beta}(s,x;k,a,b)=\left(-\frac{1}{2}\right)^{k-1}a^{-fb}f^{-s-k}\sum_{j=1}^{f}\left(\frac{\beta}{a}\right)^{jb}\chi(j)\Phi\left(\frac{\beta^{fb}}{a^{fb}},s,\frac{j+x}{f}\right).$$

**Theorem 2.10** *Let* $\chi$ *be a Dirichlet character of conductor* $f$. *Let n be a positive integer. Then we have*

$$L_{\chi,\beta}(1-n,x;k,a,b)=\frac{(-1)^k}{f}\frac{(n-1)!}{(n+k-1)!}\mathcal{Y}_{n+k-1,\chi,\beta}(x;k,a,b). \tag{16}$$

*Proof* By substituting s=1−n into (15), we get

$$L_{\chi,\beta}(1-n,x;k,a,b)=f^{n-1-k}\sum_{j=1}^{f}\left(\frac{\beta}{a}\right)^{jb}\chi(j)\zeta_{\beta^f}\left(1-n,\frac{j+x}{f};k,a^f,b\right).$$

By using Theorem 7 in [16], we get

$$L_{\chi,\beta}(1-n,x;k,a,b)$$

$$=(-1)^k \frac{(n-1)!}{(n+k-1)!} f^{n-1-k} \sum_{j=1}^{f} \left(\frac{\beta}{a}\right)^{jb} \chi(j) \mathcal{Y}_{n+k-1,\beta}\left(\frac{j+x}{f};k,a,b\right).$$

By substituting (8) into the above, we arrive at the desired result. □

**Remark 2.11** The two-variable Dirichlet $L$-function and the Dirichlet $L$-function are obvious special cases of the unified Dirichlet-type $L$-functions $L_{\chi,\beta}(s,x;k,a,b)$ defined by (13). We thus have (*cf.* [13])

$$L(s,x;\chi)=\sum_{m=0}^{\infty} \frac{\chi(m)}{(m+x)^s}$$

and

$$L(s;\chi)=\sum_{m=1}^{\infty} \frac{\chi(m)}{m^s},$$

where R(s)>1$\Re(s)>1$. By analytic continuation, this function can be extended to a meromorphic function on the whole complex plane. We have

$$L(1-n;\chi)=-\frac{B_{n,\chi}}{n},$$

where n∈Z+ and Bn,χ, the usual generalized Bernoulli number, is defined by (4). The Dirichlet $L$-function is used to prove the theorem on primes in arithmetic progressions. Dirichlet shows that L(s;χ) is non-zero at s=1. Furthermore, if $\chi$ is a principal character, then the corresponding Dirichlet $L$-function has a simple pole at s=1$s=1$ (*cf.* [6, 7, 9, 18, 24, 27, 28, 30, 31]).

# 3 APPLICATIONS

In this section, by using (16) and the following formula, which was proved by Ozden *et al.* [[16], Theorem 5, Equation (3.10)]

$$\mathcal{Y}_{n,\chi,\beta}(x;k,a,b)=\sum_{j=0}^{n}\binom{n}{j}x^{n-j}\mathcal{Y}_{j,\chi,\beta}(k,a,b), \tag{17}$$

we construct a meromorphic function involving a unified family of $L$-functions. Therefore, using (16) and (17),

$$L_{\chi,\beta}(1-n,x;k,a,b)=\frac{x^{n+k-1}}{f\prod_{l=0}^{k-1}(n+l)}\sum_{j=0}^{n+k-1}\binom{n+k-1}{j}\frac{1}{x^j}\mathcal{Y}_{j+k-1,\chi,\beta}(k,a,b).$$

From the above equation, we arrive at the following theorem.

**Theorem 3.1** *Let* $x\neq 0$. *Let* $\chi$ *be a Dirichlet character of conductor* $f$. *Then we have*

$$L_{\chi,\beta}(s,x;k,a,b)=\frac{x^{k-s}}{f\prod_{l=0}^{k-1}(s-1-l)}\sum_{j=0}^{\infty}\binom{k-s}{j}\frac{1}{x^j}\mathcal{Y}_{j+k-1,\chi,\beta}(k,a,b).$$

The function $L_{\chi,\beta}(s,x;k,a,b)$ is an analytic function at $s=0$ $s=0$. We now compute the value of this function at this point as follows:

$$L_{\chi,\beta}(0,x;k,a,b)=\frac{x^{k}}{(-1)^k f\prod_{l=0}^{k-1}(1+l)}\sum_{j=0}^{k}\binom{k}{j}\frac{1}{x^j}\mathcal{Y}_{j+k-1,\chi,\beta}(k,a,b).$$

The function $L_{\chi,\beta}(s,x;k,a,b)$ is a meromorphic function. This function has simple poles which are

$s=1,2,3,\ldots,k$.

The residues of this function at the simple poles at $s=1$ and $s=k$ are given, respectively, as follows:

$$L_{\chi,\beta}(0,x;k,a,b)=\frac{x^k}{(-1)^k f\prod_{l=0}^{k-1}(1+l)}\sum_{j=0}^{k}\binom{k}{j}\frac{1}{x^j}\mathcal{Y}_{j+k-1,\chi,\beta}(k,a,b).$$

and

$$\operatorname{Res}_{s=k}\left\{L_{\chi,\beta}(s,x;k,a,b)\right\}=\frac{\mathcal{Y}_{k-1,\chi,\beta}(k,a,b)}{f\prod_{l=0}^{k-2}(k-1-l)}.$$

**Remark 3.2** Simsek (*cf.* [20, 21]) defined a twisted two-variable *L*-function $L_{\xi,q}^{(h)}(s,x;\chi)$ as follows:

$$L_{\xi,q}^{(h)}(s,x;\chi)=\sum_{m=0}^{\infty}\frac{\chi(m)\phi_\xi(m)q^{hm}}{(x+m)^s}-\frac{\log q^h}{s-1}\sum_{m=0}^{\infty}\frac{\chi(m)\phi_\xi(m)q^{hm}}{(x+m)^{s-1}},$$

where $q\in C$ ($|q|<1$); $\xi^r=1$ ($r\in Z$); $\xi\neq 1$. Observe that if $\xi=1$, then $L_{\xi,q}^{(h)}(s,x;\chi)$ is reduced to the work of Kim [9].

Relationship between the function $L_{\chi,\beta}(s,x;k,a,b)$ and $L_{\xi,q}^{(h)}(s,x;\chi)$ is given as the following result.

**Corollary 3.3** *Let* $\chi$ *be a Dirichlet character of conductor* $f$*. Then we have*

$$L_{1,\frac{\beta^b}{a^b}}^{(b)}(s,x;\chi)=(-2)^k a^{bf} f^k\left(L_{\chi,\beta}(s,x;k,a,b)-\frac{\log q^h}{s-1}L_{\chi,\beta}(s-1,x;k,a,b)\right).$$

We conclude our present investigation by remarking that the existing literature contains several interesting generalizations and extensions of the Hurwitz-Lerch zeta function Φ(z,s,a), Hurwitz zeta function ζ(s,x) and *L*-function (*cf.* [1, 2, 3, 4, 5, 6, 7, 8, 9,10, 11, 12, 13, 14,15, 16, 17, 18, 19, 20, 21, 22, 23, 24, 25, 26, 27, 28, 29, 30]); see also the references cited in each of these earlier works.

## ACKNOWLEDGEMENTS

## Dedicated to Professor Hari M Srivastava.

Both authors are partially supported by Research Project Offices Akdeniz Universities and the Commission of Scientific Research Projects of Uludag University Project number UAP(F) 2011/38 and 2012/16. We would like to thank referees for their valuable comments.


## COMPETING INTERESTS

The authors declare that they have no competing interests.

## AUTHORS' CONTRIBUTIONS

Both authors completed the paper together. Both authors read and approved the final manuscript.

## REFERENCES


1. Ozden H: Unification of generating function of the Bernoulli, Euler and Genocchi numbers and polynomials. *AIP Conf. Proc.* 2010, 1281: 1125–1128.

2. Choi J, Srivastava HM: Some applications of the gamma and polygamma functions involving convolutions of the Rayleigh functions, multiple Euler sums and log-sine integrals. *Math. Nachr.* 2009, 282: 1709–1723.

3. Choi J, Jang DS, Srivastava HM: A generalization of the Hurwitz-Lerch zeta function. *Integral Transforms Spec. Funct.* 2008, 19: 65–79.

4. Garg M, Jain K, Srivastava HM: Some relationships between the generalized Apostol-Bernoulli polynomials and Hurwitz-Lerch zeta functions. *Integral Transforms Spec. Funct.* 2006, 17: 803–815.

5. Karande BK, Thakare NK: On the unification of Bernoulli and Bernoulli polynomials. *Indian J. Pure Appl. Math.* 1975, 6: 98–107.

6. Kim YH, Kim W, Jang LC: On the $q$-extension of Apostol-Euler numbers and polynomials. *Abstr. Appl. Anal.* 2008., 2008: Article ID 296159

7. Kim T, Rim SH, Simsek Y, Kim D: On the analogs of Bernoulli and Euler numbers, related identities and zeta and $L$-functions. *J. Korean Math. Soc.* 2008, 45: 435–453.

8. Kim T: A new approach to $q$-zeta function. *Adv. Stud. Contemp. Math.* 2005, 11: 157–162.

9. Kim T: A new approach to $p$-adic $q$-$L$-function. *Adv. Stud. Contemp. Math.* 2006, 12: 61–72.

10. Kim T: On $p$-adic $q$-$L$-functions and sums of powers. *Discrete Math.* 2002, 252: 179–187.

11. Kim T: On $p$-adic $q$-$L$-functions and sums of powers. *Discrete Math.* 2002, 252: 179–187. 10.1016/S0012-365X(01)00293-X

12. Kim T: Multiple $p$-adic $L$-function. *Russ. J. Math. Phys.* 2006, 13: 151–157.

13. Kim T, Rim SH: A note on two variable Dirichlet $L$-function. *Adv. Stud. Contemp. Math.* 2005, 10: 1–7.

14. Luo QM, Srivastava HM: Some generalizations of the Apostol-Bernoulli and Apostol-Euler polynomials. *J. Math. Anal. Appl.* 2005, 308: 290–302.

15. Ozden H, Simsek Y: A new extension of $q$ -Euler numbers and polynomials related to their interpolation functions. *Appl. Math. Lett.* 2008, 21: 934–939.

16. Ozden H, Simsek Y, Srivastava HM: A unified presentation of the generating functions of the generalized Bernoulli, Euler and Genocchi polynomials. *Comput. Math. Appl.* 2010, 60: 2779–2787. 1

17. Ráducanu D, Srivastava HM: A new class of analytic functions defined by means of a convolution operator involving the Hurwitz-Lerch zeta function. *Integral Transforms Spec. Funct.* 2007, 18: 933–943.

18. Simsek Y: $q$ -analogue of the twisted $l$ -series and $q$ -twisted Euler numbers. *J. Number Theory* 2005, 110: 267–278.

19. Simsek Y: Theorems on twisted $L$ -functions and twisted Bernoulli numbers. *Adv. Stud. Contemp. Math.* 2005, 11: 205–218.

20. Simsek Y: Twisted (h,q)$(h,q)$-Bernoulli numbers and polynomials related to twisted (h,q)$(h,q)$-zeta function and $L$ -function. *J. Math. Anal. Appl.*2006, 324: 790–804.

21. Simsek Y: Twisted $p$ -adic (h,q)-L$(h,q)$-$L$-functions. *Comput. Math. Appl.* 2010, 59: 2097–2110.

22. Simsek Y, Srivastava HM: A family of $p$ -adic twisted interpolation functions associated with the modified Bernoulli numbers. *Appl. Math. Comput.* 2010, 216: 2976–2987.

23. Srivastava HM, Ozden H, Cangul IN, Simsek Y: A unified presentation of certain meromorphic functions related to the

families of the partial zeta type functions and the $L$-functions. *Appl. Math. Comput.* 2012, 219: 3903–3913.

24. Srivastava HM, Kim T, Simsek Y: $q$-Bernoulli numbers and polynomials associated with multiple $q$-zeta functions and basic $L$-series.*Russ. J. Math. Phys.* 2005, 12: 241–268.

25. Srivastava HM, Garg M, Choudhary S: A new generalization of the Bernoulli and related polynomials. *Russ. J. Math. Phys.* 2010, 17: 251–261.

26. Srivastava HM, Garg M, Choudhary S: Some new families of generalized Euler and Genocchi polynomials. *Taiwan. J. Math.* 2011, 15: 283–305.

27. Srivastava HM, Choi J: *Series Associated with the Zeta and Related Functions*. Kluwer Academic, Dordrecht; 2001.

28. Srivastava HM, Choi J: *Zeta and q-Zeta Functions and Associated Series and Integrals*. Elsevier, Amsterdam; 2012.

29. Apostol TM: On the Lerch zeta function. *Pac. J. Math.* 1951, 1: 161–167.

30. Whittaker ET, Watson GN: *A Course of Modern Analysis. An Introduction to the General Theory Infinite Processes and of Analytic Functions: With an Account of the Principal Transcendental Functions*. 4th edition. Cambridge University Press, Cambridge; 1962.

31. Yildirim CY: Zeros of derivatives of Dirichlet $L$-function. *Turk. J. Math.* 1996, 20: 521–534

4

# ON SOME APPLICATIONS OF INTEGRAL *p*-ADIC HODGE THEORY TO GALOIS REPRESENTATIONS

**Go Yamashita a,1, Seidai Yasuda b,2**

a Toyota Central R&D Labs. Inc., 41-1, Yokomichi, Nagakute, Aichi, 480-1192, Japan b Department of Mathematics, Graduate School of Science, Osaka University, Osaka, 560-0043, Japan

## ABSTRACT

We explicitly construct an analytic family of $n$-dimensional crystalline representations by using integral $p$-adic Hodge theory. This is a generalization of results by Berger, Li, and Zhu and by Dousmanis. We show, by using Kisin's method, that the part of a universal deformation ring related to the above constructions is connected. From this we obtain an explicitly described subclass of potentially diagonalizable representations in the sense of Barnet-Lamb, Gee, Geraghty and Taylor. This yields automorphy lifting theorem and potential automorphy theorem, in which the condition at $p$ is weakened.

## KEYWORDS

Automorphy (modularity) lifting theorems; Integral ***p***-adic Hodge theory; Wach module; Moduli of Wach modules

# 1. INTRODUCTION

Kisin introduced a new technique on automorphy (modularity) lifting theorems. The technique is to study local deformation rings by using a moduli of semi-linear algebraic objects (SS-modules, and Wach modules) in the integral $p$ -adic Hodge theory, and to show $R^{red}$=TRred=T by using the properties of local deformation rings. In the $GL_2$GL2 case, this method had an important application to Serre's conjecture.

On the other hand, Clozel, Harris and Taylor proved some automorphy lifting theorems for unitary groups in [CHT]. Their method has been improved in [G], [T] and [BLGHT] (in [T], Taylor used Kisin's technique so-called in "ℓ≠pℓ≠p" case) and has produced an important application to Sato–Tate conjecture [HSBT] and [BLGHT]. However, the automorphy lifting theorem has not been established in the best possible form, in the sense that they require certain local conditions of the involved Galois representations. A main difficulty in removing such local conditions lies in the study of local deformation rings in the case where the coefficient field has the same characteristic $p$ as the residue field of the considered local field (so-called in "ℓ=pℓ=p" case).

Currently, one of the strongest available statements is given in [BLGGT], where an automorphy lifting theorem has been established under some mild condition in the case where the Galois representation, restricted to the decomposition group at each prime dividing $p$, is "potentially diagonalizable" in the sense of [BLGGT]. However, the notion of potential diagonalizability is defined in terms of a universal deformation ring whose structure has not yet been well-understood. Owing to this, potential diagonalizability is in some sense like a tautological synonym for the local condition with which we can manage to show an automorphy of a global Galois representation by the currently available methods, and it seems that we do not have a description of the potentially diagonalizable representations which is concrete enough for applications.

In this paper we construct, for any integer $n \geq 2n \geq 2$ and for any finite unramified extension *K* of $Q_pQp$, an explicitly described subclass of potentially diagonalizable *n*-dimensional *p* -adic representations (depending on a choice of some extra parameters $w, \underline{r}, \underline{\alpha}$w,$\underline{r}$,$\underline{\alpha}$ which we introduce in Definition 2.7) of the absolute Galois group $G_KGK$ of *K*, by using the method of Berger, Li and Zhu [BLZ], and Kisin [K1] and [K2] (see Proposition 5.2). More precisely,

- First, we construct an explicit analytic family of Wach modules by the method of Berger, Li and Zhu (see Proposition 3.12 and Corollary 3.13).

- Next, we show a property on the connected components of local deformation rings by a method similar to that of Kisin (see Proposition 4.2).

The above extra parameters define a *p* -adic representation of $G_KGK$ which is a direct sum of representations induced from characters of open subgroups of $G_KGK$ corresponding to finite unramified extensions of *K*. The subclass consists of the representations which are *p*-adically "near" this direct sum. We call the members of the subclass the *absolutely nearly semi-induced* representations. From our construction, combined with the result in [BLGGT], we deduce an automorphy lifting and a potentially automorphy result for an essentially conjugate self-dual *p*-adic representation *r* of the absolute Galois group of a CM field *F*, such that *F* is unramified at *p* and for each prime *v* of *F* dividing *p*, the restriction of *r* to the decomposition group at *v* is absolutely nearly semi-induced or a successive extension of one-dimensional representations (under some other mild conditions for *r*). See Corollary 5.3 and Corollary 5.4 for the precise statements. We emphasize the following two things:

- The notion of the potentially diagonalizability in [BLGGT] means the representations which are **geometrically** near to the induced representations. On the other hand, the notion of absolutely nearly semi-induced representations in this paper means the representations which are $p$-**adically** near to (the direct sums of) the induced representations.

- Our results are quantitative. We gave a $p$-adic estimate in which the automorphic lifting machinery works (see also Definition 3.11).

The extra parameter $\underline{r}\underline{r}$ stands for the Hodge–Tate weights and we can construct the subclass of representations starting from *any* fixed Hodge–Tate weights (however, when we apply [BLGGT] after that, we have to assume that it has distinct Hodge–Tate weights). However our results still remain partial and unfortunately they are not sufficient to cover all the $n$ -dimensional crystalline representations of $G_K$GK with fixed Hodge–Tate weights, except for the case where the Hodge–Tate weights are in an interval of length p−1, and for some very limited case where n=2n=2 and the difference of any two Hodge–Tate weights is in {−p,0,p}. It seems difficult[3] to extend our method, even when n=2n=2 and the extra parameter $w$ is not the identity, to cover the case where one of the eigenvalues of crystalline Frobenius is not close to zero. We note that, for such cases, a partial result has been obtained in [K2] when $K=Q_p$ and n=2, based on the computation by Berger and Breuil of the reductions of some crystalline representations using the compatibility of the mod pmod p and the $p$ -adic local Langlands correspondences for $GL_2(Q_p)$ (see also Remark 3.14).

After a submission of our paper, a referee kindly informed us of the references [D], [Z], [B3] and [GL] whose subjects are related to that in our paper. All of these papers are concerned with the computation of the reduction of a crystalline representation $V$ of $G_K$GK with $K/Q_p$ unramified. We briefly explain their relations to our paper. The main results of [D] and the calculations of the reductions in [D] are exclusively for the case where $V$ is two dimensional. However some two dimensional cases not covered in our paper are treated there. Moreover in Section 4 of [D], Dousmanis gives a criterion for a pair (P,G) of matrices to be an approximation of the pair of the matrices of $\varphi$ and $\gamma$ on some family of Wach modules. The approximation treats not only the case of the direct sum of the induced representations like us, but also the case of more general ones. However he does not explicitly construct the family of Wach modules when the rank is greater than two. In this sense, our paper is a generalization of [D]. In [Z] and [B3] they proved, among some other results, that any crystalline representation $V'V'$ in an explicitly given $p$-adic neighbourhood of $V$ has the same reduction as $V$. However they do not give an explicit computation of the reduction. (They also treat other topics. However, they have little relation to our paper.) In [GL], they treat only the case where the length of the Hodge filtration is less than or equal to $p-1p-1$. In this sense, our paper is a generalization of [GL] as well.

Let us give a brief summary of the structure of the paper. In Section 2, we construct a certain family of Galois representations by using the theory of Wach modules developed in [B1]. Our family is a generalization of the family constructed in [BLZ]. In Section 3, we introduce the notion of nearly semi-induced representations. In Section 4, we recall the deformation rings introduced by Kisin [K2] and [K3] and check that the family of nearly semi-induced representations are contained in one connected component of the local deformation space. In Section 5, we describe some applications to automorphy lifting and potential automorphy.

# 2. A FAMILY OF WACH MODULES

Let $n\geq 1n\geq 1$ be a positive integer. In this section, we construct an analytic family of $n$-dimensional crystalline representations by using the theory of Wach modules, and calculate its mod $p$ reduction. This is a generalization of the work by Berger, Li and Zhu [BLZ].

## 2.1. Notation

In this section, we fix a rational prime $p$. Let $K$ be a finite unramified extension of $Q_pQp$. Let $val:K\to Z\cup\{\infty\}val:$ denote the normalized valuation on $K$, and let $W\subset K$ be the valuation ring of $K$ with respect to the valuation val. Let $\sigma$ denote the Frobenius on $W$ , that is, the unique automorphism of the $\mathbb{Z}_p$-algebra $W$ which induces the $p$ -power Frobenius automorphism on W/pWW/pW. Since K=W[1/p], the automorphism $\sigma$ on $W$ is uniquely extended to the automorphism of $K$, which we denote by the same symbol $\sigma$ by abuse of notation. Fix an algebraic closure $\overline{K}$ of $K$ and put $G_K := \mathrm{Gal}(\overline{K}/K)$.. We put $\Gamma_K$:=Gal(K($\mu_{p^\infty}$)/K) and let $\chi : \Gamma_K \xrightarrow{\sim} \mathbb{Z}_p^\times$ denote the $p$-adic cyclotomic character.

Let us consider the ring K[[π]] of formal power series with coefficients in $K$ with respect to the formal variable $\pi$ . Let φ:K[[π]]→K[[π]] denote the $\sigma$-semi-linear, $\pi$-adically continuous endomorphism of the $K$ -algebra K[[π]] which sends $\pi$ to $(1+\pi)^p-1$. For γ∈$\Gamma_K$, let γ:K[[π]]→K[[π]] denote the $K$-linear, $\pi$-adically continuous endomorphism of the $K$ -algebra K[[π]] which sends $\pi$ to

$$(1+\pi)^{\chi(\gamma)} - 1 = \sum_{n=1}^{\infty} \frac{\prod_{i=0}^{n-1}(\chi(\gamma) - i) \cdot \pi^n}{n!}.$$

This gives an action of the group $\Gamma_K$ on the $K$ -algebra K[[π]].

## 2.2. Review of the theory of Wach modules and Wach lattices

Let $\mathbf{A}_K^+ := W[[\pi]] \subset K[[\pi]]$ denote the subring of formal power series with coefficients in $W$ and put $\mathbf{B}_K^+ := \mathbf{A}_K^+[1/p]$. The rings $\mathbf{A}_K^+$ and $\mathbf{B}_K^+$ are subrings of K[[π]]K[[π]] which are stable under the actions of $\varphi$ and $\Gamma$ . We recall the definition of Wach modules and Wach lattices over $\mathbf{A}_K^+$ and over $\mathbf{B}_K^+$ (cf. [B1] and [BLZ]).

Let

$$\mathbf{A}_K := W[[\pi]][1/\pi]^\wedge = \left\{ \sum_{i=-\infty}^{+\infty} a_i\pi^i \;\middle|\; a_i \in W, \mathrm{val}(a_i) \to \infty \text{ as } i \to -\infty \right\}$$

denote the $p$ -adic completion of the ring $\mathbf{A}_K^+[1/\pi]$. The ring $\mathbf{A}_K$AK is a discrete valuation ring and $p$ is a uniformizer of $\mathbf{A}_K$. Let $\mathbf{B}_K:=\mathbf{A}_K[1/p]$ be the field of fractions of $\mathbf{A}_K$. The endomorphism $\varphi$ of $\mathbf{A}_K^+$ has a unique extension to a $\sigma$ -semilinear endomorphism of $\mathbf{A}_K$, which we also denote by $\varphi$, such that the latter $\varphi$ is continuous with respect to the $p$ -adic topology. The action of the group $\Gamma_K$ΓK on $\mathbf{A}_K^+$ has a unique extension to the action of $\Gamma_K$ on $\mathbf{A}_K$ such that for each $\gamma\in\Gamma_K$, the automorphism $\gamma:\mathbf{A}_K\to\mathbf{A}_K$ of $\mathbf{A}_K$ given by the action of $\gamma$ is continuous with respect to the $p$-adic topology.

Let $R$ be a finite $Z_p$Zp-algebra. We extend the endomorphism $\varphi$ and the action of $\Gamma_K$ΓK to $(\mathbf{A}_K)_R:=R\otimes_{Z_p}\mathbf{A}_K$ by $R$ -linearity. A $(\varphi,\Gamma_K)$ *-module over* $(\mathbf{A}_K)_R$ is a pair $D=(D,\varphi_D)$ of a finitely generated free $(\mathbf{A}_K)_R$ -module $D$ and a $\varphi$ -semi-linear endomorphism $\varphi_D$φD of $D$, equipped with a continuous (with respect to the $p$ -adic topology) semi-linear action of the group $\Gamma_K$ such that for each $\gamma\in\Gamma_K$ the semi-linear automorphism of $D$ given by the action of $\gamma$ commutes with the endomorphism $\varphi_D$φD. We say that a $(\varphi,\Gamma_K)$ -module $D$ over $(\mathbf{A}_K)_R$ is *étale* if the subset $\varphi_D(D)\subset D$ generates $D$ over $(\mathbf{A}_K)_R$.

Let $E$ is a finite extension of $Q_p$ and let $O_E$ denote its ring of integers. We extend the endomorphism $\varphi$ and the action of $\Gamma_K$ to $(\mathbf{B}_K)_E:=E\otimes_{Z_p}\mathbf{A}_K$ by $E$ -linearity. A $(\varphi,\Gamma_K)$ *-module over* $(\mathbf{B}_K)_E$ is a pair $D=(D,\varphi_D)$ of a finitely generated free $(\mathbf{B}_K)_E$ -module $D$ and a $\varphi$ -semi-linear endomorphism $\varphi_D$ of $D$, equipped with a continuous (with respect to the $p$ -adic topology) semi-linear action of the group $\Gamma_K$ such that for each $\gamma\in\Gamma_K$ the semi-linear automorphism of $D$ given by the action of $\gamma$ commutes with the endomorphism $\varphi_D$ We say that a $(\varphi,\Gamma_K)$ -module $D$ over $(\mathbf{B}_K)_E$ is *étale* if there exists an $(\mathbf{A}_K)_{O_E}$ -lattice $D_0\subset D$ stable under the actions of $\varphi$ and $\Gamma_K$ such that $D_0$ is an étale $(\varphi,\Gamma_K)$ -module over $(\mathbf{A}_K)_{O_E}$.

For a finite $Z_p$ -algebra $R$ (resp. a finite extension $R$ of $Q_p$), a *free R-representation* of $G_K$ is a finitely generated free $R$-module $T$ equipped with a continuous, $R$ -linear action $\rho:G_K\to \mathrm{End}_R(T)$ρ: where we endow $\mathrm{End}_R(T)$ with the $p$-adic topology. In [F], Fontaine constructs a functor $\mathbf{D}$ from the category of free $Z_p$Zp-representations of $G_K$ to the category of étale $(\varphi,\Gamma_K)$ -modules over $\mathbf{A}_K$, and shows that the functor **D** gives an equivalence of the categories. By extending this functor by *R*-linearity, we have a functor, also denoted by **D**, which gives an equivalence between the category of free *R* -representations of $G_K$GK to the category of étale $(\varphi,\Gamma_K$-modules over $(\mathbf{A}_K)_R$ (resp. $(\mathbf{B}_K)_R$(BK)R).

From now on we fix a finite extension *E* of $Q_p$ and let $O_E$ denote its ring of integers. Let $m_E$mE denote the maximal ideal of $O_E$, and let F be the residue field $O_E/m_E$. Below we use the subscripts $(-)_E$ and $(-)_{OE}$ which respectively mean the operations $E\otimes_{Qp}(-)$ and $O_E\otimes_{Zp}(-)$OE. Since *W* is unramified over $Z_p$, the $O_E$-algebra $W_{OE}=O_E\otimes_{Zp}W$ is equal to the direct product $W_{OE}=\prod_{\xi\in\Xi}W_\xi$, where for each $\xi\in\Xi$, the $O_E$-algebra $W_\xi$ is the ring of integers of a finite unramified extension of *E* , and the index set $\Xi=\Xi_E$ is a finite set whose cardinality *c* is equal to the greatest common divisor of $[F:F_p]$ and $[K:Q_p]$[K:Qp]. The Frobenius automorphism $\sigma$ $\sigma = \mathrm{id}_{O_E}\otimes\sigma$ of $W_{O_E}$ of $W_{OE}$WOE induces a cyclic permutation $\sigma_\Xi:\Xi\to\Xi$ of order *c*, and a $\sigma$ -linear isomorphism $W_\xi \xrightarrow{\cong} W_{\sigma_\Xi(\xi)}$ of $W_{O_E}$ of $W_{OE}$ -algebras for each $\xi\in\Xi$. For a $W_{OE}$ -algebra *R* and for $\xi\in\Xi$, we put $R_\xi=W_\xi\otimes_{WOE}R$ so that $R=\prod_{\xi\in\Xi}R_\xi$. The endomorphism $\varphi$ on $\varphi$ on $(\mathbf{A}_K^+)_{O_E}$ induces an isomorphism $\varphi:(\mathbf{A}_K^+)_{O_E,\xi}\xrightarrow{\cong}(\mathbf{A}_K^+)_{O_E,\sigma_\Xi(\xi)}$ for each $\xi\in\Xi$

In [B1] (see also [BLZ]), Berger gives a criterion for a free *E* - representation of $G_K$GK to be crystalline which is described purely in terms of its associated $(\varphi,\Gamma_K)$-module. Let us recall the statement of the criterion. We put $q:=\varphi(\pi)/\pi=\sum_{i=1}^{p}\binom{p}{i}\pi^{i-1}$ which we regard an element in $(\mathbf{B}_K^+)_E=(\mathbf{A}_K^+[1/p])_E$. For an invertible element $x=(x_\xi)_{\xi\in\Xi}$ in $(\mathbf{B}_K)_E=\prod_{\xi\in\Xi}(\mathbf{B}_K)_{E,\xi}$ and for a tuple $\underline{a}=(a\xi)$ of integers indexed by $\xi\in\Xi$ we let $x\underline{a}$ denote the element $(x_\xi^{a_\xi})_{\xi\in\Xi}$ in $(\mathbf{B}_K)_E$. Let *V* be a free *E* - representation of $G_K$GK of dimension $d:=\dim_E V$d:=dimEV and let $\underline{a}=(a_\xi)_{\xi\in\Xi}$ and $\underline{b}=(b_\xi)_{\xi\in\Xi}$ be two tuples of integers indexed by $\xi\in\Xi$ such that $a_\xi\le b_\xi$ holds for every $\xi\in\Xi$. Then a free *E*-representation *V* of $G_K$GK is crystalline with Hodge–Tate weights in the tuple $[\underline{a},\underline{b}]:=([a_\xi,b_\xi])_{\xi\in\Xi}$ of intervals if and only if the associated $(\varphi,\Gamma_K)$-

module $(\mathbf{D}(V), \varphi_{\mathbf{D}(V)})$ over $(\mathbf{B}_K)_E$ has a $(\mathbf{B}_K^+)_E$-submodule $\mathbf{N}(V) \subset \mathbf{D}(V)$ satisfying the following conditions:

(1)The module **N**(V) is free of rank *d* over $(\mathbf{B}_K^+)_E$.

(2)The action of $\Gamma_K$ on **D**(V) preserves **N**(V).

(3)The action of $\Gamma_K$ΓK on **N**(V)N(V) induces a trivial action on **N**(V)/π**N**(V).

(4)We have $\varphi_{\mathbf{D}(V)}(\pi^{\underline{b}}\mathbf{N}(V)) \subset \pi^{\underline{b}}\mathbf{N}(V)$ and the quotient $\pi^{\underline{b}}\mathbf{N}(V)/\varphi^*_{\mathbf{D}(V)}(\pi^{\underline{b}}\mathbf{N}(V))$ is killed by $q^{\underline{b}-\underline{a}}$. Here $\varphi^*_{\mathbf{D}(V)}(\pi^{\underline{b}}\mathbf{N}(V))$φD(V)?(πbN(V)) denotes the $(\mathbf{B}_K^+)_E$ -submodule of $\pi^{\underline{b}}\mathbf{N}(V)$ generated by the subset $\varphi_{\mathbf{D}(V)}(\pi^{\underline{b}}\mathbf{N}(V)) \subset \pi^{\underline{b}}\mathbf{N}(V)$.

Moreover, if *V* is a free *E* -representation of $G_K$GK which is crystalline with Hodge–Tate weights in [a,b], then a $(\mathbf{B}_K^+)_E$ -submodule **N**(V)⊂**D**(V)N(V)⊂D(V) satisfying the above conditions is unique and the condition (4) holds even when we replace a,b with any tuples of integers $\underline{a}' = (a'_\xi)$, $\underline{b}' = (b'_\xi)$ with $a'_\xi \le b'_\xi$ for every ξ∈Ξ such that the Hodge–Tate weights of *V* are in the tuple [a′,b′] of intervals. Hence for a crystalline free *E*-representation *V* of $G_K$GK, the notation **N**(V)N(V) makes sense even if we do not specify a range of its Hodge–Tate weights.

When *V* is crystalline, Berger [B1] also gives an isomorphism $\mathbf{N}(V)/\pi\mathbf{N}(V) \xrightarrow{\cong}$ Dcrys(V)E of filtered $\varphi$-modules with coefficient in *E* . Here we endow **N**(V)/π with the structure of a filtered $\varphi$-module in the following way. The endomorphism $\varphi$ of **N**(V)/π**N**(V) is the one induced from the endomorphism $(q/p)^b\varphi$ of **N**(V) for a sufficiently large integer *b* . We filter **N**(V) by putting $\mathrm{Fil}^i$**N**(V):={x∈**N**(V)|$\varphi_{\mathbf{D}(V)}$(x)∈$q^i$**N**(V)} for i∈Z, and endow the *E* -vector space **N**(V)/π**N**(V) with the filtration induced by the filtration $(\mathrm{Fil}^i\mathbf{N}(V))_{i\in Z}$ N(V).

A *Wach module* over $(\mathbf{B}_K^+)_E$ with Hodge–Tate weights in [$\underline{a}$,$\underline{b}$] is a free $(\mathbf{B}_K^+)_E$ -module $N$ of finite rank equipped with a semi-linear action $\Gamma_K \to \mathrm{Aut}_E(N)$ΓK→AutE(N) of $\Gamma_K$ΓK which is continuous with respect to the $p$ -adic topology of $\mathrm{Aut}_E(N)$AutE(N) and with a $\varphi$ -semi-linear homomorphism $\varphi_N : (\mathbf{B}_K^+)_E[1/\pi] \otimes_{(\mathbf{B}_K^+)_E} N \to (\mathbf{B}_K^+)_E[1/\varphi(\pi)] \otimes_{(\mathbf{B}_K^+)_E} N$. which commutes with the action of $\Gamma_K$ΓK such that the following conditions hold:

(1)The action of $\Gamma_K$ΓK on $N$ induces a trivial action on $N/\pi$.

(2)We have $\varphi_N(\pi^{\underline{b}}N)$ and the quotient $\pi^{\underline{b}}N/\varphi_N^*(\pi^{\underline{b}}N)$ is killed by $q^{\underline{b}-\underline{a}}$qb−a. Here $\varphi_N^*(\pi^{\underline{b}}N)$ denotes the $(\mathbf{B}_K^+)_E$-submodule of $\pi^{\underline{b}}N$ generated by the subset $\varphi_N(\pi^{\underline{b}}N) \subset \pi^{\underline{b}}N$.

If $(N,\varphi_N)$ is a Wach module over $(\mathbf{B}_K^+)_E$ with Hodge–Tate weights in [$\underline{a}$,$\underline{b}$], then the action of $\Gamma_K$ΓK on $N$ (resp. the homomorphism $\varphi_N$φN) induces a semi-linear action of $\Gamma_K$ΓK on the $(\mathbf{B}_K)_E$(BK)E-module $D(N) = (\mathbf{B}_K)_E \otimes_{(\mathbf{B}_K^+)_E} N$ (resp. a $\varphi$ -semi-linear endomorphism $\varphi_{D(N)}$ of D(N), and the $(\mathbf{B}_K)_E$(BK)E-modules together with the endomorphism $\varphi_{D(N)}$ and the action of $\Gamma_K$ΓK form an étale $(\varphi,\Gamma_K)$-module over $(\mathbf{B}_K)_E$.

Let $V$ be a crystalline free $E$ -representation of $G_K$ with Hodge–Tate weights in a tuple [$\underline{a}$,$\underline{b}$] of intervals. If $T$ is a $G_K$-stable $O_E$OE-lattice in $V$ , then $\mathbf{N}(T):=\mathbf{D}(T)\cap\mathbf{N}(V)$ is an $(\mathbf{A}_K^+)_{O_E}$-lattice in $\mathbf{N}(V)$. The map $T\mapsto\mathbf{N}(T)$ gives a bijection between the $G_K$-stable $O_E$OE-lattices in $V$ and the $(\mathbf{A}_K^+)_{O_E}$ -submodules $N$ of $\mathbf{N}(V)$ satisfying the following conditions:

(1)The $(\mathbf{A}_K^+)_{O_E}$ (AK+)OE-module $N$ is free of rank $d=\dim_E$V.

(2)The action of $\Gamma_K$ΓK on $\mathbf{N}(V)$N(V) preserves $N$.

(3)We have $\varphi_{\mathbf{D}(V)}(\pi^{\underline{b}}N) \subset \pi^{\underline{b}}N$ and the quotient $\pi^{\underline{b}}N/\varphi_{\mathbf{D}(V)}^*(\pi^{\underline{b}}N)$ is killed by $q^{\underline{b}-\underline{a}}$. Here $\varphi_{\mathbf{D}(V)}^*(\pi^{\underline{b}}N)$ denotes the $(\mathbf{B}_K^+)_E$ -submodule of $\pi^{\underline{b}}N$ generated by the subset $\varphi_{\mathbf{D}(V)}(\pi^{\underline{b}}N) \subset \pi^{\underline{b}}N$.

We call an $(\mathbf{A}_K^+)_{\mathcal{O}_E}$ -submodule *N* of **N**(V)N(V) satisfying the above conditions a *Wach lattice* in **N**(V)N(V).

## 2.3. Some elementary properties of $\mathbf{A}_{\mathbb{Q}_p}^+$

In this subsection, we assume that $K=Q_p$. For later use we state the following two statements as lemmas. Proofs of the statements are easy and are left to the reader.

**Lemma 2.1.** *For any* $\gamma \in \Gamma_{Qp}$, *the element* γ(π)/πγ(π)/π*is invertible in* $\mathbf{A}_{\mathbb{Q}_p}^+$ □

**Lemma 2.2.** *The sequence* $\{\varphi^i(1+\pi)\}_{i\geq 0}$ *converges to* 1 *in* $\mathbf{A}_{\mathbb{Q}_p}^+$ *with respect to the* (p,π)*-adic topology on* $\mathbf{A}_{\mathbb{Q}_p}^+$ □

**Lemma 2.3.** *The sequence* $\{\varphi^i(q)\}_{i\geq 0}$$\{\varphi i(q)\}i\geq 0$*converges to p in* $\mathbf{A}_{\mathbb{Q}_p}^+$AQp+*with respect to the* (p,π)(p,π)*-adic topology.*

**Proof.**

Since $\varphi^i(q) = \sum_{j=0}^{p-1} \varphi^i((1+\pi)^j)$, the claim follows from Lemma 2.2. □

We define a subring $R_{\mathbb{Q}_p}$ of $\mathbb{Q}_p[[\pi]]$ as follows:

$$R_{\mathbb{Q}_p} := \left\{ \sum_{i\geq 0} c_i \pi^i \in \mathbb{Q}_p[[\pi]] \;\middle|\; \mathrm{val}(c_i) + \frac{i}{p-1} \geq 0 \text{ for all } i \geq 0 \right\}$$

The subring $R_{Qp}$ is stable under $\varphi$ and the action of $\Gamma_K$. We endow the ring $R_{Qp}$ with the linear topology such that the set

$$\{p^i R_{\mathbb{Q}_p} + (R_{\mathbb{Q}_p} \cap \pi^j \mathbb{Q}_p[[\pi]]) \mid i,j \in \mathbb{Z}_{\geq 0}\}$$

of ideals in $R_{Qp}$ forms a fundamental system of neighbourhoods of 0. We easily see that the ring $R_{Qp}$ is complete with respect to this topology and that the element q/p is invertible in $R_{Qp}$.

Lemma 2.3 has the following corollary:

**Corollary 2.4.** *The sequence of invertible elements* $\{\varphi^i(q)/p\}_{i\geq 0}$ *converges to* 1 *in* $R_{Qp}$.

The following lemma can be verified by direct calculation:

**Lemma 2.5.**

(1) *The ring* $\mathbf{A}^+_{\mathbb{Q}_p}$ AQp+*is a closed subring in* $R_{Qp}$.

(2) *The topology on* $\mathbf{A}^+_{\mathbb{Q}_p}$ *induced from that on* $R_{Qp}$ *by the inclusion* $\mathbf{A}^+_{\mathbb{Q}_p} \subset R_{\mathbb{Q}_p}$ *coincides the* (p,π)(p,π)*-adic topology on* $\mathbf{A}^+_{\mathbb{Q}_p}$.

By combining Corollary 2.4 with Lemma 2.1, we get the following:

**Corollary 2.6.** *For any* $\gamma \in \Gamma_{Qp}$, *the sequence* $\{\varphi^i(q/\gamma(q))\}_{i\geq 0}$ *converges to* 1 *in* $\mathbf{A}^+_{\mathbb{Q}_p}$ *with respect to the* (p,π) *-adic topology.*

For a map $f: Z_{\geq 0} \to Z$, we define an element $\lambda(f) \in R_{Qp}$ to be the infinite product

$$\lambda(f) := \prod_{i\geq 0} \left(\varphi^i(q)/p\right)^{f(i)}.$$

It follows from Corollary 2.4 that this infinite product converges in $R_{Qp}$RQp.

It is straightforward to check the element $\lambda(f) \in R_{Qp}$ has the following properties:

We have $\lambda(f) \in R^{\times}_{\mathbb{Q}_p}$

- For two maps $f_1, f_2: Z_{\geq 0} \to Z$, we have $\lambda(f_1+f_2)=\lambda(f_1)\lambda(f_2)$.
- For a map $f: Z_{\geq 0} \to Z$, let $Sf: Z_{\geq 0} \to Z$ denote the map

$$Sf(i) := \begin{cases} 0 & \text{if } i = 0, \\ f(i-1) & \text{if } i \geq 1. \end{cases}$$

Then we have $\varphi(\lambda(f))=\lambda(Sf)$ .

## 2.4. The matrix $P((X_{i,j,\theta})_{i,j[1,n],\theta\Theta})$

Let us return to the case for a general finite unramified extension $K$ of $Q_p$. Fix an integer $n\geq1$ and put $[1,n]:=\{1,\ldots,n\}$ . For a ring $A$ , let $M_n(A)$ denote the ring of $n$-by-$n$ matrices with entries in $A$. Let $E$ be a finite extension of $Q_p$ and let $O_E$ denote its ring of integers. We fix $\xi_0\in\Xi$ and put $\xi_h = \sigma_{\Xi}^h(\xi_0)$ for $h = 1,\ldots,c-1$. We fix an $O_E$ -basis $(e_\theta)_{\theta\in\Theta}$ of $W_{\xi0}$, where the index set $\Theta$ is a finite set of cardinality $[K:Q_p]/c$. Let $\sigma^{(c)} = (\sigma^c_{\theta,\theta'})_{\theta,\theta'\in\Theta}$ denote the matrix of the $O_E$-linear automorphism $\sigma^c\sigma c$ of $W_{\xi0}$ with respect to the basis $(e_\theta)_{\theta\in\Theta}$.

We let

$$(\mathbf{A}_K^+)_{O_E}[[\underline{X}]] := (\mathbf{A}_K^+)_{O_E}[[X_{i,j,\theta}(1 \leq i,j \leq n, \theta \in \Theta)]]$$

denote the ring of formal power series with $n^2\cdot[K:Q_p]/c$ variables over $(\mathbf{A}_K^+)_{O_E}$(AK+)OE. For $\xi\in\Xi$, we put $(\mathbf{A}_K^+)_{O_E,\xi}[[\underline{X}]] = (\mathbf{A}_K^+)_{O_E,\xi}[[X_{i,j,\theta}(1 \leq i,j \leq n, \theta \in \Theta)]]$ so that $(\mathbf{A}_K^+)_{O_E}[[\underline{X}]] = \prod_{\xi\in\Xi}(\mathbf{A}_K^+)_{O_E,\xi}[[\underline{X}]]$.

**Definition 2.7.**

We define a set $\mathrm{Data}_{K,E,n}$DataK,E,n to be the set of triples $(w,\underline{r},\underline{\alpha})$ of an element $w$ in the $n$ -th symmetric group Aut([1,n])Aut([1,n]), a tuple $\underline{r}=(r_{i,\xi})_{(i,\xi)\in[1,n]\times\Xi}$ of integers $r_{i,\xi}\in Zr$ indexed by the pairs $(i,\xi)\in[1,n]\times\Xi$, and a tuple $\underline{\alpha} = (\alpha_{i,\xi})_{(i,\xi)\in[1,n]\times\Xi}$ of elements $\alpha_{i,\xi} \in W_\xi^\times$. indexed by the pairs $(i,\xi)\in[1,n]\times\Xi$. We call an element in $\mathrm{Data}_{K,E,n}$ *a basic datum of degree n* . We say that a basic datum $(w,(r_{i,\xi}),(\alpha_{i,\xi}))$ is *effective* if $r_{i,\xi}$ is non-negative for every $(i,\xi)\in[1,n]\times\Xi$. We say that a basic datum $(w,(r_{i,\xi}),(\alpha_{i,\xi}))$ has *distinct Hodge–Tate weights* if the integers $r_{1,\xi},\ldots,r_{n,\xi}$ are distinct for every $\xi\in\Xi$.

Fix an effective basic datum $(w,\underline{r},\underline{\alpha}) = (w,(r_{i,\xi}),(\alpha_{i,\xi})) \in \mathrm{Data}_{K,E,n}$ of degree $n$. The goal of this subsection is to introduce a matrix $P(\underline{X}) \in M_n((\mathbf{A}_K^+)_{\mathcal{O}_E}[[\underline{X}]])$, which depends on the basic data $(w,\underline{r},\underline{\alpha})$.

We put $r_{\xi,\min}=\min_{i\in[1,n]} r_{i,\xi}$ and $r_{\xi,\max}=\max_{i\in[1,n]} r_{i,\xi}$. Let $\tilde{w}$ denote the automorphism of the set $[1,n]\times\Xi$ which sends $(i,\xi)\in[1,n]\times\Xi$ to $(i,\sigma_\Xi^{-1}(\xi))$ if $\xi\neq\xi_0$ and which sends $(i,\xi_0)$ to $(w(i),\sigma_\Xi^{-1}(\xi_0))$. Let us consider the following three conditions on the data $w$ and $(r_{i,\xi},\alpha_{i,\xi})_{(i,\xi)\in[1,n]\times\Xi}$:

- There exists a pair $(i,\xi)\in[1,n]\times\Xi$ such that $r_{\tilde{w}^k(i,\xi)} = r_{\sigma_\Xi^{-k}(\xi),\max}$ holds for any $k\geq 0$.
- There exists a pair $(i,\xi)\in[1,n]\times\Xi$ such that $r_{\tilde{w}^k(i,\xi)} = r_{\sigma_\Xi^{-k}(\xi),\min}$ holds for any $k\geq 0$.
- $r_{\xi,\max}-r_{\xi,\min}$ is independent of $\xi\in\Xi$

We define $\delta$ to be $\delta:=1$ if all of the conditions hold, and $\delta:=0$ otherwise. For example, if $\tilde{w}$ has no fixed points and if the integers $(r_{i,\xi})_{(i,\xi)\in[1,n]\times\Xi}$ are different with each others, then neither of conditions holds and hence $\delta=0$ in this case. We put

$$r := \delta + \max_{\xi\in\Xi}(r_{\xi,\max} - r_{\xi,\min}). \tag{2.1}$$

We need some more notation before introducing the matrix $P(\underline{X})$. For $\xi\in\Xi$ and for $f_\xi \in (\mathbf{A}_K^+)_{\mathcal{O}_E,\xi}[[\underline{X}]]$, let $\varphi'(f_\xi) \in (\mathbf{A}_K^+)_{\mathcal{O}_E,\sigma_\Xi(\xi)}[[\underline{X}]]$ denote the formal power series obtained by applying $\varphi$ on each coefficient of $f_\xi$. For $f=(f_\xi)_{\xi\in\Xi} \in (\mathbf{A}_K^+)_{\mathcal{O}_E}[[\underline{X}]]$ let $\varphi(f)$ denote the element in $(\mathbf{A}_K^+)_{\mathcal{O}_E}[[\underline{X}]]$ whose component at $\xi\in\Xi$ is equal to $\varphi'(f_{\sigma_\Xi^{-1}(\xi)})$ for $\xi\neq\xi_0$ and whose component at $\xi_0$ is equal to

$$\varphi'(f_{\sigma_{\Xi}^{-1}(\xi_0)})\left(\left(\sum_{\theta'\in\Theta}\sigma^{(c)}_{\theta,\theta'}X_{i,j,\theta'}\right)_{i,j,\theta}\right)\in(\mathbf{A}_K^+)_{O_E,\xi_0}[[\underline{X}]].$$

For $f\in(\mathbf{A}_K^+)_{O_E,\xi}[[\underline{X}]]$ and for $\gamma\in\Gamma_K$, let $\gamma(f)$ denote the formal power series obtained by applying $\gamma$ on each coefficient of $f$. For a matrix $M$ in $M_n(K[[\pi]])$Mn(K[[π]]) or in $M_n((\mathbf{A}_K^+)_{O_E}[[\underline{X}]])$, let $\varphi(M)$ (resp. $\gamma(M)$) denote the matrix obtained by applying $\varphi$ (resp. $\gamma$) on the entries of $M$.

Let $1_n$1n be the identity $n$-by-$n$ matrix, and let $C$ be the permutation matrix corresponding to $w^{-1}$w−1, that is, $C$ is the $n$-by-$n$ matrix whose $(i,j)$-entry is $\delta_{w(i),j}$ where $\delta_{*,*}$ is the Kronecker's delta.

For each $(i,\xi)\in[1,n]\times\Xi$, let $f_{i,\xi}:\mathbb{Z}_{\geq 0}\to\mathbb{Z}_{\geq 0}$ denote the map which sends $j\in\mathbb{Z}_{\geq 0}$

$$f_{i,\xi}(j):=r_{\widetilde{w}^j(i,\xi)}.$$

For $i,j\in[1,n]$ and for $\xi\in\Xi$, we write

$$\lambda(f_{i,\xi})/\lambda(f_{j,\xi})=\sum_{k\geq 0}z'^{(k)}_{i,j,\xi}\pi^k.$$

Let $m$ denote the smallest non-negative integer such that $p^m z'^{(k)}_{i,j,\xi}\in\mathbb{Z}_p$ for every $i,j\in[1,n]$i,j∈[1,n], for every $\xi\in\Xi$, and for every $0\leq k\leq r-1$. Since $\lambda(f_{i,\xi})/\lambda(f_{j,\xi})\in R_{\mathbb{Q}_p}$, it follows from the definition of $R_{\mathbb{Q}_p}$ that $m\leq(r-1)/(p-1)$.

**Example 1.** One of the simplest cases where we cannot apply Fontaine–Laffaille theory is $r=p$. In this case, if $r_{i,\xi}=0$ or $p$ for any $i\in[1,n]$, $\xi\in\Xi$, then we have $m=0$. However, unfortunately for $n\geq 3$, this condition ($r_{i,\xi}=0$ri,ξ=0, or $p$ for any $i\in[1,n]$I, $\xi\in\Xi$ cannot hold with distinct Hodge–Tate weights. So, if $r=p$ with distinct Hodge–Tate weights, and $n\geq 3$, then we have $m=1$. For $n=2$, it can hold with distinct Hodge–Tate weights and we have m=0m=0 (see [BLZ, Remark 4.1.2, 1]).

We write $z_{i,j,\xi}^{(k)} = p^m z_{i,j,\xi}'^{(k)}$ (k) and put

$$z_{i,j,\xi} := \sum_{k=0}^{r-1} z_{i,j,\xi}^{(k)} \pi^k,$$

which is, by the definition of $m$ , an element in $\mathbf{A}_{\mathbb{Q}_p}^+$

For ξ∈Ξ, let $Z_\xi$ denote the matrix in Mn(Wξ[[π]][[X̲]]) whose (i,j)(i,j)-entry is $z_{i,j,\xi} \cdot \sum_{\theta \in \Theta} X_{i,j,\theta}$. For ξ∈Ξ, define a matrix Pξ(X̲)∈Mn(Wξ[[π]][[X̲]]) to be

$$P_\xi(\underline{X}) := \begin{cases} (1_n + Z_{\xi_0}) \cdot \mathrm{diag}(\alpha_{1,\xi_0} q^{r_{1,\xi_0}}, \ldots, \alpha_{n,\xi_0} q^{r_{n,\xi_0}}) \cdot C, & \text{if } \xi = \xi_0, \\ (1_n + Z_\xi) \cdot \mathrm{diag}(\alpha_{1,\xi} q^{r_{1,\xi}}, \ldots, \alpha_{n,\xi} q^{r_{n,\xi}}), & \text{if } \xi \neq \xi_0 \end{cases}$$

and put $P(\underline{X}) := (P_\xi(\underline{X}))_{\xi \in \Xi} \in M_n((\mathbf{A}_K^+)_{\mathcal{O}_E}[[\underline{X}]])$.Here dia g$(\alpha_{1,\xi} q^{r_{1,\xi}}, \ldots, \alpha_{n,\xi} q^{r_{n,\xi}})$ denotes the diagonal matrix whose diagonal entries are $\alpha_{1,\xi} q^{r_{1,\xi}}, \ldots, \alpha_{n,\xi} q^{r_{n,\xi}}$.

## 2.5. The matrix $G_\gamma(r)$

For ξ∈, we let $G_\xi \in M_n(R_{\mathbb{Q}_p})$ denote the matrix

$$G_\xi := \mathrm{diag}\big(\lambda(f_{1,\xi}), \ldots, \lambda(f_{n,\xi})\big).$$

It follows from the properties described in Section 2.3 that the matrix $G_\xi$Gξ is invertible (as a matrix with entries in $R_{\mathbb{Q}_p}$).

**Lemma 2.8.**

*We have*

$$G_\xi = \mathrm{diag}\big((q/p)^{r_{1,\xi}}, \ldots, (q/p)^{r_{n,\xi}}\big) \cdot \varphi(G_{\sigma_\Xi^{-1}(\xi)})$$

*for* $\xi \neq \xi_0$ *and*

$$G_{\xi_0} = \mathrm{diag}\big((q/p)^{r_{1,\xi_0}}, \ldots, (q/p)^{r_{n,\xi_0}}\big) C\varphi(G_{\sigma_\Xi^{-1}(\xi_0)})C^{-1}.$$

**Proof.** Let $i \in [1,n]$. We have $f_{\tilde{w}(i,\xi)}(j-1) = f_{i,\xi}(j)$ for $j \geq 1$. Hence the equality

$$\lambda(f_{i,\xi}) = (q/p)^{r_{i,\xi}} \varphi\big(\lambda(f_{\tilde{w}(i,\xi)})\big)$$

holds from which the claim immediately follows for $\xi \neq \xi_0$ The claim for $\xi = \xi_0$ follows from the equality

$$C\varphi(G_{\sigma_\Xi^{-1}(\xi_0)})C^{-1} = \mathrm{diag}\big(\varphi(\lambda(f_{w(1),\sigma_\Xi^{-1}(\xi_0)})), \ldots, \varphi(\lambda(f_{w(n),\sigma_\Xi^{-1}(\xi_0)}))\big).$$

□

For $\xi \in \Xi$ and for $\gamma \in \Gamma_K$, we let $G^{(r)}_{\gamma,\xi}$ denote the *n*-by-*n* matrix

$$G_\xi \gamma(G_\xi)^{-1} = \mathrm{diag}\big(\lambda(f_{1,\xi})/\gamma(\lambda(f_{1,\xi})), \ldots, \lambda(f_{n,\xi})/\gamma(\lambda(f_{n,\xi}))\big).$$

It follows from Corollary 2.6 that the matrix $G^{(r)}_{\gamma,\xi}$ is an invertible matrix with entries in $\mathbf{A}^+_{\mathbb{Q}_p}$. We put $G^{(r)}_\gamma = (G^{(r)}_{\gamma,\xi})_{\xi\in\Xi}$ which we regard an element in $M_n((\mathbf{A}^+_K)_{O_E}) = \prod_{\xi\in\Xi} M_n((\mathbf{A}^+_K)_{O_E,\xi})$.

**Lemma 2.9.** *We have*

$$P(0)\varphi(G^{(r)}_\gamma) = G^{(r)}_\gamma \gamma(P(0)).$$

**Proof.** For a matrix $M \in M_n((\mathbf{A}^+_K)_{O_E}) = \prod_{\xi\in\Xi} M_n((\mathbf{A}^+_K)_{O_E,\xi})$ and for $\xi \in \Xi$, let $M_\xi$ denote the component of *M* at *ξ*. By Lemma 2.8, we have the equalities

$$G^{(r)}_{\gamma,\xi} = \mathrm{diag}\big((q/\gamma(q))^{r_{1,\xi}}, \ldots, (q/\gamma(q))^{r_{n,\xi}}\big) \cdot \varphi(G^{(r)}_{\gamma})_{\xi}$$

for $\xi \neq \xi_0$ and

$$G^{(r)}_{\gamma,\xi_0} = \mathrm{diag}\big((q/\gamma(q))^{r_{1,\xi_0}}, \ldots, (q/\gamma(q))^{r_{n,\xi_0}}\big) \cdot C\varphi(G^{(r)}_{\gamma})_{\xi_0} C^{-1}.$$

On the other hand, the equalities

$$\mathrm{diag}\big((q/\gamma(q))^{r_{1,\xi}}, \ldots, (q/\gamma(q))^{r_{n,\xi}}\big) = \gamma\big(P(0)_{\xi}\big)^{-1} P(0)_{\xi}$$

for $\xi \neq \xi_0$ and

$$\mathrm{diag}\big((q/\gamma(q))^{r_{1,\xi_0}}, \ldots, (q/\gamma(q))^{r_{n,\xi_0}}\big) = C\gamma\big(P(0)_{\xi_0}\big)^{-1} P(0)_{\xi_0} C^{-1}$$

follow from the definition of P(0). By combining these four equalities, we have

$$G^{(r)}_{\gamma} = \widetilde{C}\gamma\big(P(0)\big)^{-1} P(0)\widetilde{C}^{-1} \cdot \widetilde{C}\varphi(G^{(r)}_{\gamma})\widetilde{C}^{-1}$$

where $\widetilde{C} \in M_n(W_{\mathcal{O}_E}) = \prod_{\xi \in \Xi} M_n(W_{\xi})$ denotes the matrix whose component at $\xi_0$ is equal to $C$ and whose other components are equal to the identity matrix. The three matrices $\widetilde{C}\gamma(P(0))^{-1}$, $P(0)\widetilde{C}^{-1}$, $\widetilde{C}\varphi(G^{(r)}_{\gamma})\widetilde{C}^{-1}$ commute with each other since they are diagonal matrices. Hence the right hand side of the above formula can be rewritten as

$$P(0)\widetilde{C}^{-1} \cdot \widetilde{C}\varphi(G^{(r)}_{\gamma})\widetilde{C}^{-1} \cdot \widetilde{C}\gamma\big(P(0)\big)^{-1} = P(0)\varphi(G^{(r)}_{\gamma})\gamma\big(P(0)\big)^{-1}.$$

Thus we have $G^{(r)}_{\gamma} = P(0)\varphi(G^{(r)}_{\gamma})\gamma(P(0))^{-1}$, which proves the claim. □

**Proposition 2.10.**

*The entries of the matrix*

$$P(\underline{X})\varphi(G^{(r)}_{\gamma})\gamma\big(P(\underline{X})\big)^{-1} - G^{(r)}_{\gamma}$$

*are in* $\pi^r(\mathbf{A}_K^+)_{O_E}[[\underline{X}]]$.

**Proof.**

By the definition of $P(\underline{X})$, we have

$$P(\underline{X})P(0)^{-1} = 1_n + Z, \tag{2.2}$$

where $Z \in M_n((\mathbf{A}_K^+)_{O_E}[[\underline{X}]]) = \prod_{\xi\in\Xi} M_n((\mathbf{A}_K^+)_{O_E,\xi}[[\underline{X}]])$ denotes the matrix whose component at $\xi$ is equal to $Z_\xi$ for every $\xi \in \Xi$. The matrix $1_n+Z$1n+Z is congruent to $1_n$1n modulo the maximal ideal of $(\mathbf{A}_K^+)_{O_E}[[\underline{X}]]$(AK+)OE[[X]]. Hence by (2.2), the matrix $\gamma(P(\underline{X}))\gamma(P(0))^{-1}$ is invertible in $M_n((\mathbf{A}_K^+)_{O_E}[[\underline{X}]])$. Thus to prove the claim, it suffices to show that the entries of the matrix

$$Y := \left(P(\underline{X})\varphi(G_\gamma^{(r)})\gamma(P(\underline{X}))^{-1} - G_\gamma^{(r)}\right)\cdot\gamma(P(\underline{X}))\gamma(P(0))^{-1}$$

are in $\pi^r(\mathbf{A}_K^+)_{O_E}[[\underline{X}]]$

By (2.2), we have

$$\begin{aligned} Y &= \left(P(\underline{X})\varphi(G_\gamma^{(r)}) - G_\gamma^{(r)}\gamma(P(\underline{X}))\right)\cdot\gamma(P(0))^{-1} \\ &= P(\underline{X})\varphi(G_\gamma^{(r)})\gamma(P(0))^{-1} - G_\gamma^{(r)}\gamma(1_n+Z) \\ &= (1_n+Z)P(0)\varphi(G_\gamma^{(r)})\gamma(P(0))^{-1} - G_\gamma^{(r)}\gamma(1_n+Z). \end{aligned}$$

We also have $P(0)\varphi(G_\gamma^{(r)})\gamma(P(0))^{-1} = G_\gamma^{(r)}$, by using Lemma 2.9. Thus we have

$$\begin{aligned} Y &= (1_n+Z)G_\gamma^{(r)} - G_\gamma^{(r)}\gamma(1_n+Z) \\ &= ZG_\gamma^{(r)} - G_\gamma^{(r)}\gamma(Z). \end{aligned}$$

For h=0,...,c−1h=0,...,c−1, it follows from the definitions of Gγ(r) and $Z$ that the component at $\xi_h$ of the (i,j) -entry of $Y$ is

$$z_{i,j,\xi_h}\lambda(f_{j,\xi_h})/\gamma(\lambda(f_{j,\xi_h})) - \gamma(z_{i,j,\xi_h})\lambda(f_{i,\xi_h})/\gamma(\lambda(f_{i,\xi_h}))$$

multiplied by $\sum_{\theta\in\Theta} X_{i,j,\theta}$ Let $K_{\xi h}$ denote the field of fractions of $W_{\xi h}$. Then by the definition of $z_{i,j,\xi h}$, the element $z_{i,j,\xi h}-p^m\lambda(f_{i,\xi h})/\lambda(f_{j,\xi h})$ is in $\pi^r K_{\xi h}[[\pi]]$. Therefore, the component at $\xi_h$ of the (i,j) -entry of $Y$ is an element of $\pi^r K_{\xi_h}[[\pi]] \cap W_{\xi_h}[[\pi]] = \pi^r W_{\xi_h}[[\pi]]$ multiplied by $\sum_{\theta\in\Theta} X_{i,j,\theta}$. In particular, the (i,j)-entry of $Y$ lies in $\pi^r(\mathbf{A}_K^+)_{O_E}[[\underline{X}]]$. We are done. □

## 2.6. The matrix $G_\gamma$

**Proposition 2.11.** *For each $\gamma\in\Gamma_K$, there exists a unique matrix $G_\gamma \in M_n((\mathbf{A}_K^+)_{O_E}[[\underline{X}]])$ satisfying the following conditions:*

$$G_\gamma \equiv G_\gamma^{(r)} \bmod \pi^r(\mathbf{A}_K^+)_{O_E}[[\underline{X}]], \text{ and}$$

$$P(\underline{X})\varphi(G_\gamma) = G_\gamma\gamma(P(\underline{X})).$$

Before going to the proof, we will introduce some notation. We put $P_0 := P(\underline{X}) \bmod \pi(\mathbf{A}_K^+)_{O_E}[[\underline{X}]]$ and $Z_0 := Z \bmod \pi(\mathbf{A}_K^+)_{O_E}[[\underline{X}]]$. They are *n*-by-*n* matrices with entries in $(\mathbf{A}_K^+)_{O_E}/(\pi) = W_{O_E}[[\underline{X}]]$For $\xi\in\Xi$, we let $D_{0,\xi}$ denote the diagonal matrix $\mathrm{diag}(\alpha_1 p^{r_{1,\xi}},\ldots,\alpha_n p^{r_{n,\xi}})\in M_n(W_\xi)$ and put $D_0=(D_{0,\xi})_{\xi\in\Xi}\in M_n(W_{OE})$. It follows from the definition of $P_0$ that the equality

$$P_0 = (1_n + Z_0)D_0\widetilde{C} \qquad (2.3)$$

holds, which shows that the matrix $P_0P0$ is invertible as a matrix with entries in $W_{OE}[[\pi]][1/p]$.

For $\xi\in\Xi$ and for $f0,\xi\in W\xi[[\underline{X}]]$, let $\sigma'(f_{0,\xi}) \in W_{\sigma_\Xi(\xi)}[[\underline{X}]]$ denote the formal power series obtained by applying $\sigma$ on each coefficient of $f_{0,\xi}$. For $f_0 = (f_{0,\xi})_{\xi\in\Xi} \in W_{O_E}[[\underline{X}]]$, let $\sigma(f_0)$ denote the element in $W_{O_E}[[\underline{X}]]$ whose component at $\xi\in\Xi$ is equal to $\sigma'(f_{0,\sigma_\Xi^{-1}(\xi)})$ for $\xi\neq\xi_0$ and whose component at $\xi_0\xi0$ is equal to

$$\sigma'(f_{0,\sigma_{\Xi}^{-1}(\xi_0)})\left(\left(\sum_{\theta'\in\Theta}\sigma_{\theta,\theta'}^{(c)}X_{i,j,\theta'}\right)_{i,j,\theta}\right)\in W_{\xi_0}[[\underline{X}]].$$

For $M_0\in M_n(W_{O_E}[[\underline{X}]])$, let $\sigma(M_0)$ denote the matrix obtained by applying $\sigma$ on the entries of $M_0$.

Let $M_0$M0 be an element in $W_{O_E}[[\underline{X}]]$ (resp. in $M_n(W_{O_E}[[\underline{X}]])$. For any $M\in(\mathbf{A}_K^+)_{O_E}[[\underline{X}]]$ (resp. for any $M\in M_n((\mathbf{A}_K^+)_{O_E}[[\underline{X}]])$ satisfying $M_0\equiv M \bmod \pi(\mathbf{A}_K^+)_{O_E}[[\underline{X}]]$, the following properties are easily checked:

$\sigma(M_0)\equiv\varphi(M) \bmod \pi(\mathbf{A}_K^+)_{O_E}[[\underline{X}]]$, and

$\gamma(M_0)\equiv\gamma(M) \bmod \pi(\mathbf{A}_K^+)_{O_E}[[\underline{X}]]$ for any $\gamma\in\Gamma_K$.

For $M_0\in M_n(W_{O_E}[[\underline{X}]])$, put

$$\Psi(M_0):=p^rP_0\sigma(M_0)P_0^{-1}.$$

By (2.1) and (2.3), the component $\Psi(M_0)_\xi$Ψ(M0)ξ of $\Psi(M_0)$Ψ(M0) at $\xi\in\Xi$ξ∈Ξ is equal to

$$\Psi(M_0)_\xi=p^{\delta_\xi}(1_n+Z_0)(p^{-r_{\xi,\min}}D_0)\widetilde{C}\sigma(M_0)\widetilde{C}^{-1}(p^{r_{\xi,\max}}D_0^{-1})(1_n+Z_0)^{-1},\quad(2.4)$$

where $\delta_\xi=r-(r_{\xi,\max}-r_{\xi,\min})$. Thus $\Psi(M_0)$ lies in $M_n(W_{O_E}[[\underline{X}]])$, and any entries of $\Psi(M_0)$ are divisible by $p$ if $\delta=1$. Let m be the Jacobson radical of $W_{O_E}[[\underline{X}]]$ It is clear that the operation $M_0\mapsto\Psi(M_0)$ is $\sigma$-semilinear. Hence $\Psi$ induces a semilinear operation $\Psi_0:=\Psi \bmod \mathfrak{m}$ Ψ0:=Ψmodm on $M_n(W_{O_E}[[\underline{X}]]/\mathfrak{m})=M_n(W_{O_E}\otimes_{O_E}\mathbb{F})$.

The following is a key lemma for the proof of Proposition 2.11:

**Lemma 2.12.**

(1)*The operation* $\Psi_0$Ψ0*is nilpotent, i.e., there exists an integer* i≥1*such that* Ψ0i=0. *Furthermore, we can take such an i satisfying* i≤nc.

(2)*The operation $\Psi$ is* mm-*adically nilpotent in the following sense: for any integer* $k\geq 1k\geq 1$, *there exists an integer i, such that the entries of* $\Psi^i(M_0)\Psi i(M0)$*are in* $m^k mk$*for any* $M_0 \in M_n(W_{O_E}[[\underline{X}]])$.

**Proof.** First we prove (1). Suppose that $\delta=1$. Since $\Psi_0=0$ by (2.4), the claim is clear in this case. Let us assume $\delta=0$ . For $i,j\in[1,n]i$, and for $\xi\in\Xi$ \let $E_{i,j,\xi} \in M_n(W_{O_E}[[\underline{X}]])$ denote the matrix such that the component at $\xi$ of the $(i,j)(i,j)$-entry is equal to 1 and that the other components and other entries are equal to 0. Let $\Xi'\subset\Xi$ denote the subset of the elements $\xi\in\Xi$ satisfying $\delta_\xi=0\delta\xi=0$. Let $I$ denote the subset of $[1,n]\times\Xi$ which consists of the elements $(i,\xi)\in[1,n]\times\Xi'$ satisfying $r_{i,\xi}=r_{\xi,\min}$, and let $J$ denote the subset of $[1,n]\times\Xi$ which consists of the elements $(i,\xi)\in[1,n]\times\Xi'$ satisfying $r_{i,\xi}=r_{\xi,\max}$.

We consider the reductions modulo mm of the both sides of (2.4). We have $Z_0 \equiv 0 \bmod \mathfrak{m}$, and $(p^{-r_{\xi,\min}}D_{0,\xi})_{\xi\in\Xi} \bmod \mathfrak{m}$ is a diagonal matrix such that the component at $\xi$ of the $i$ -th diagonal entry is invertible if $(i,\xi)\in I(i,\xi)\in I$ and is zero otherwise. On the other hand, $(p^{r_{\xi,\max}}D_{0,\xi}^{-1})_{\xi\in\Xi} \bmod \mathfrak{m}$ is a diagonal matrix such that the component at $\xi$ of the $i$ -th diagonal entry is invertible if $(i,\xi)\in J$ and is zero otherwise. Thus it follows from (2.4) that $\Psi_0(E_{i,j,\xi})$ is a multiple of $E_{i,j,\sigma_\Xi(\xi)}$ for $\sigma_\Xi(\xi)\neq\xi_0$, and of $E_{w^{-1}(i),w^{-1}(j),\xi_0}$ for by a constant in $W[[\underline{X}]]/\mathfrak{m}W[[\underline{X}]]/m$, and $\Psi_0(E_{i,j,\xi})=0$ unless $\widetilde{w}^{-1}(i,\xi)\in I$ and $\widetilde{w}^{-1}(j,\xi)\in J$. (Note that $\widetilde{w}^{-1}(i,\xi)$ is equal to $(i,\sigma_\Xi(\xi))$ for $\sigma_\Xi(\xi)\neq\xi_0$, and to $(w^{-1}(i),\xi_0)$ for $\sigma_\Xi(\xi)=\xi_0$ Since we have assumed $\delta=0$, there exists, for any $i,j\in[1,n]$ and for any $\xi\in\Xi$, an integer $k\leq nc$ such that $\widetilde{w}^{-k}(i,\xi)\notin I$ or $\widetilde{w}^{-k}(j,\xi)\notin J$. Hence we have $\Psi_0^{nc}(E_{i,j,\xi})=0\Psi 0nc(Ei,j,\xi)=0$ for any $i,j\in[1,n]i,j\in[1,n]$ and for any $\xi\in\Xi$ Therefore $\Psi_0^{nc}=0$ which prove the claim (1).

We prove the claim (2) by induction on $k$ . If $k=1k=1$, the claim follows from (1). Suppose that $k\geq 2k\geq 2$, and assume that the statement holds for $k-1k-1$. By the induction hypothesis, there exists an integer $i'i'$ such that the entries of $\Psi 0i'(M0)$ are in $m^{k-1}$ for any $M_0 \in M_n(W_{O_E}[[\underline{X}]])$. $M0\in Mn(WOE[[\underline{X}]])$. Thus there exist an integer $\ell\geq 0\ell\geq 0$, elements $x_1,\ldots,x_\ell\in m^{\ell-1}x1,\ldots,x\ell\in m\ell-1$, and

$N_1,\ldots,N_\ell \in M_n(W_{O_E}[[\underline{X}]])$ such that $\Psi_0^{i'}(M_0)$ can be written as a finite sum

$$\Psi_0^{i'}(M_0) = \sum_{j=1}^{\ell} x_j N_j.$$

Then, for i:=i′+nci:=i′+nc, we obtain

$$\Psi_0^{i}(M_0) = \sum_{j=1}^{\ell} \sigma^{nc}(x_j)\Psi_0^{nc}(N_j).$$

Hence it follows from (1) that the entries of $\Psi_0^{i'}(M_0)$ are in m$^k$mk. Thus the statement holds for *k*. □

**Proof of Proposition 2.11.** First we prove the uniqueness of $G_\gamma$Gγ. Suppose that two matrices $G_\gamma$Gγ and $G'_\gamma$, with $G_\gamma \neq G'_\gamma$, satisfy the conditions of Proposition 2.11. Let *k* be the maximum non-negative integer satisfying $G_\gamma \equiv G'_\gamma \bmod \pi^k(\mathbf{A}_K^+)_{O_E}[[\underline{X}]]$. We have k≥rk≥r, since $G_\gamma \equiv G'_\gamma \bmod \pi^r(\mathbf{A}_K^+)_{O_E}[[\underline{X}]]$ by assumption. Put $G_\gamma - G'_\gamma = \pi^k H$, where $H \in M_n((\mathbf{A}_K^+)_{O_E}[[\underline{X}]])$. By assumption, we have $P(\underline{X})\varphi(\pi^k H) = \pi^k H\gamma(P(\underline{X}))$, that is,

$$q^k P(\underline{X})\varphi(H) = H\gamma(P(\underline{X})). \tag{2.5}$$

Put $H_0 := H \bmod \pi(\mathbf{A}_K^+)_{O_E}[[\underline{X}]]$ Then $H_0$H0 is in $M_n((\mathbf{A}_K^+)_{O_E}[[\underline{X}]]/(\pi)) = M_n(W_{O_E}[[\underline{X}]])$, and $H_0 \neq 0$ by the definition of *k*. By (2.5), we have $p^k P_0 \sigma(H_0) = H_0 P_0$. Thus we have

$$H_0 = p^{k-r}\Psi(H_0). \tag{2.6}$$

By applying (2.6) repeatedly, we have

$$H_0 = p^{i(k-r)}\Psi^i(H_0)$$

for any i≥1. It follows from Lemma 2.12 that the entries of the right hand side of the above formula converge to 0 with respect to the mm-adic topology of $W_{\mathcal{O}_E}[[\underline{X}]]$. Hence we have $H_0$=0H0=0, which gives a contradiction. This proves that the matrix $G_\gamma$Gγ satisfying the conditions of Proposition 2.11 is unique if it exists.

Next we will show the existence of $G_\gamma$Gγ. It suffices to construct $G_\gamma^k \in M_n((\mathbf{A}_K^+)_{\mathcal{O}_E}[[\underline{X}]])$ for each k≥rk≥r satisfying

- $G_\gamma^{(k)} \equiv G_\gamma^{(r)} \bmod \pi^r(\mathbf{A}_K^+)_{\mathcal{O}_E}[[\underline{X}]]$,
- $G_\gamma^{(k)} \equiv G_\gamma^{(k-1)} \bmod \pi^{k-1}(\mathbf{A}_K^+)_{\mathcal{O}_E}[[\underline{X}]]$ for $k > r$, and
- $P(\underline{X})\varphi(G_\gamma^{(k)})\gamma(P(\underline{X}))^{-1} \equiv G_\gamma^{(k)} \bmod \pi^k(\mathbf{A}_K^+)_{\mathcal{O}_E}[[\underline{X}]]$,

since we can define $G_\gamma$Gγ as the limit of $G_\gamma^{(k)}$. We construct such a $G_\gamma^{(k)}$ by induction on $k$ . For k=rk=r, we put $G_\gamma^{(k)} := G_\gamma^{(r)}$. We assume that k>rk>r and that $G_\gamma^{(k-1)}$ is constructed. To construct $G_\gamma^{(k)}$, it is enough to find a matrix $H \in M_n((\mathbf{A}_K^+)_{\mathcal{O}_E}[[\underline{X}]])$ satisfying

$$P(\underline{X})\varphi(G_\gamma^{(k-1)} + \pi^{k-1}H)\gamma(P(\underline{X}))^{-1} \equiv G_\gamma^{(k-1)} + \pi^{k-1}H \bmod \pi^k(\mathbf{A}_K^+)_{\mathcal{O}_E}[[\underline{X}]]. \tag{2.7}$$

By assumption, there exists $Y \in M_n((\mathbf{A}_K^+)_{\mathcal{O}_E}[[\underline{X}]])$ satisfying $P(\underline{X})\varphi(G_\gamma^{(k-1)})\gamma(P(\underline{X}))^{-1} - G_\gamma^{(k-1)} = \pi^{k-1}Y$. Put $H_1 := H \bmod \pi(\mathbf{A}_K^+)_{\mathcal{O}_E}[[\underline{X}]]$, $Y_0 := Y \bmod \pi(\mathbf{A}_K^+)_{\mathcal{O}_E}[[\underline{X}]]$ The equality (2.7) is equivalent to

$$H_1 - p^{k-1}P_0\varphi(H_1)P_0^{-1} = Y_0,$$

that is,

$$H_1 - p^{k-r-1}\Psi(H_1) = Y_0. \tag{2.8}$$

Let us consider an infinite sum

$$H_1 := \sum_{i \geq 0} p^{i(k-r-1)} \Psi^i(Y_0).$$

Lemma 2.12 shows that this infinite sum mm-adically converges, and satisfies (2.8). Take an arbitrary lift $H \in M_n((\mathbf{A}_K^+)_{\mathcal{O}_E}[[\underline{X}]])$ of $H_1$H1. Then $H$ satisfies (2.7). Therefore the matrix $G_\gamma^{(k)} := G_\gamma^{(k-1)} + \pi^{k-1} H$ satisfies the conditions listed in the previous paragraph. This completes the proof of the existence of $G_\gamma$. □

# 3. THE REPRESENTATION OF TYPE $(W,(R_{I,\Xi}),(A_{I,\Xi}))$(W,(RI,Ξ),(AI,Ξ))

We put

$$\text{Tuple} := \prod_{i,j \in [1,n], \theta \in \Theta} \mathfrak{m}_E.$$

For $\underline{a} = (a_{i,j,\theta}) \in \text{Tuple}$, we define an $O_E$OE-algebra homomorphism $\mathrm{ev}_{\underline{a}} : (\mathbf{A}_K^+)_{\mathcal{O}_E}[[\underline{X}]] \to (\mathbf{A}_K^+)_{\mathcal{O}_E}$ to be the homomorphism which sends $f(\underline{X}) = (f_\xi(\underline{X})) \in (\mathbf{A}_K^+)_{\mathcal{O}_E}[[\underline{X}]]$ to $(f_{\xi_h}(a_{i,j,\theta}\sigma^h(e_\theta)))_{0 \leq h \leq c-1} \in (\mathbf{A}_K^+)_{\mathcal{O}_E}$ It is easy to check that the homomorphism $\mathrm{ev}_{\underline{a}}$ is compatible with the endomorphisms $\varphi$ of $(\mathbf{A}_K^+)_{\mathcal{O}_E}[[\underline{X}]]$ and $(\mathbf{A}_K^+)_{\mathcal{O}_E}$, and with the actions of $\Gamma_K$ on and $(\mathbf{A}_K^+)_{\mathcal{O}_E}$.

Let $n \geq 1$n≥1 and let $(w, \underline{r}, \underline{\alpha}) \in \mathrm{Data}_{K,E,n}$(w,r,α)∈DataK,E,n be an effective basic datum of degree $n$. We put $r_{\xi,\min} = \min_{i \in [1,n]} r_{i,\xi}$ and $r_{\xi,\max} = \max_{i, \in [1,n] \times \Xi}$,∈[1,n]×Ξri,Ξ. Let $P(\underline{X})$ and $G_\gamma$ for $\gamma \in \Gamma_K$ be as in the previous section. For $\underline{a} \in \text{Tuple}$a∈Tuple, we put $P_{\underline{a}} := \mathrm{ev}_{\underline{a}}(P(\underline{X}))$ and $G_{\gamma,\underline{a}} := \mathrm{ev}_{\underline{a}}(G_\gamma)$ Let $N_{\underline{a},\mathcal{O}_E} = (\mathbf{A}_K^+)_{\mathcal{O}_E}^{\oplus n}$ be a free $(\mathbf{A}_K^+)_{\mathcal{O}_E}$ module of rank $n$. Let $\varphi_{N,\underline{a}}$φN,a be the $\varphi$ -semi-linear endomorphism of $N_{\underline{a},\mathcal{O}_E}$ whose matrix with respect to the standard basis is equal to $P_{\underline{a}}$. For $\gamma \in \Gamma_K$, let $\gamma_{\underline{a}}$ be the $\gamma$ -semi-linear endomorphism of $N_{\underline{a},\mathcal{O}_E}$ ,OE whose matrix with respect to the standard basis is equal to $G_{\gamma,\underline{a}}$.

The map $\gamma \mapsto \gamma_{\underline{a}}$ gives a semi-linear action of the group $\Gamma_K$ on $N_{\underline{a},\mathcal{O}_E}$ ,OE. We put $N_{\underline{a}} = E \otimes_{\mathcal{O}_E} N_{\underline{a},\mathcal{O}_E}$ which is a free $(\mathbf{B}_K^+)_E$(BK+)E-module of rank $n$ . We extend the endomorphism $\varphi_{N,\underline{a}}$φN,a̲ of $N_{\underline{a},\mathcal{O}_E}$Na̲,OE (resp. the action of $\Gamma_K$ΓK on $N_{\underline{a},\mathcal{O}_E}$Na̲,OE) to that of $N_{\underline{a}}$ (resp. that on $N_{\underline{a}}$) by $E$ -linearity. The endomorphism $\varphi_{N,\underline{a}}$ a̲ on $N_{\underline{a}}$ induces a $\varphi$ -semi-linear homomorphism $(\mathbf{B}_K^+)_E[1/\pi] \otimes_{(\mathbf{B}_K^+)_E} N_{\underline{a}} \to (\mathbf{B}_K^+)_E[1/\varphi(\pi)] \otimes_{(\mathbf{B}_K^+)_E} N_{\underline{a}}$ which we denote by the same symbol φN,a̲ by abuse of notation. The following lemma is easily checked:

**Lemma 3.1.**

*The* $(\mathbf{B}_K^+)_E$ *-module* $N_{\underline{a}}$ *together with the endomorphism* $\varphi_{N,\underline{a}}$ *and the action of* $\Gamma_K$ΓK*form a Wach module over* $(\mathbf{B}_K^+)_E$ E*with the Hodge–Tate weights in the tuple* $([-r_{\sigma\Xi(\xi),\max}, -r_{\sigma\Xi(\xi),\min}])_{\xi\in\Xi}$ *of intervals and the* $(\mathbf{A}_K^+)_{\mathcal{O}_E}$ *-module* $N_{\underline{a},\mathcal{O}_E}$ *is a Wach lattice of the étale* $(\varphi,\Gamma_K)$ *-module* $N_{\underline{a}} \otimes_{(\mathbf{B}_K^+)_E} (\mathbf{B}_K)_E$.

Let $V_{\underline{a}}$ be the free $E$ -representation of $G_K$GK corresponding to the étale $(\varphi,\Gamma_K)$-module $N_{\underline{a}} \otimes_{(\mathbf{B}_K^+)_E} (\mathbf{B}_K)_E$ over $(\mathbf{B}_K)_E$. Then $V$ is crystalline with the Hodge–Tate weights in $([-r_{\sigma\Xi(\xi),\max}, -r_{\sigma\Xi(\xi),\min}])_{\xi\in\Xi}$ Let $T_{\underline{a}} \subset V_{\underline{a}}$ be the $G_K$GK-stable $O_E$OE-lattice corresponding to the Wach lattice $N_{\underline{a},\mathcal{O}_E}$. The following lemma follows from the argument in [BLZ, Theorem 4.1.1].

**Lemma 3.2.** *For* $\underline{a} \in \mathrm{Tuple}$, *there is a canonical isomorphism* $\mathbb{F} \otimes_{\mathcal{O}_E} T_{\underline{a}} \cong \mathbb{F} \otimes_{\mathcal{O}_E} T_{\underline{0}}$ *of free* FF*-representation of* $G_K$ *where* $\underline{0} = (0)_{i,j\in[1,n],\theta\in\Theta} \in \mathrm{Tuple}_0$.

We need the following lemma.

**Lemma 3.3.** *Let* ξ∈ *and let* $s \geq 1$ *be a positive integer. Let* $x$ *be an element in* $(\mathbf{B}_K^+)_{E,\xi}$,ξ*satisfying* $\varphi(x) \in q^s (\mathbf{B}_K^+)_{E,\xi}$. *Then* $x \in \pi^s (\mathbf{B}_K^+)_{E,\xi}$.

**Proof.** Take a primitive $p$ -th root $\zeta_p$ζp of unity in an algebraic closure $\overline{K_\xi}$ of $K_\xi = E \otimes_{O_E} W_\xi$ and let $W'_\xi = W_\xi[\zeta_p]$. Let $g$ be the E′-linear (π,p)-adically continuous automorphism of $W'_\xi[[\pi]]$ which sends $\pi$ to $\zeta_p(1+\pi)-1$ζp(1+π)−1.

Since $g((1+\pi)^p)=(1+\pi)^p$, we have $g\circ\varphi=\varphi g\circ\varphi=\varphi$ on $W'_\xi[[\pi]]$. Hence $\varphi(x)=g(\varphi(x))\in g(q)^s W'_\xi[[\pi]]$. Since $g(q)=\pi q/(\zeta_p(1+\pi)-1)$ is divisible by $\pi$, we have $\varphi(x)\in(\mathbf{B}_K^+)_{E,\xi}\cap\pi^s W'_\xi[[\pi]]=\pi^s(\mathbf{B}_K^+)_{E,\xi}$ E,ξ. This shows $x\in\pi^s(\mathbf{B}_K^+)_{E,\xi}$x∈πs(BK+)E,ξ. □

**Corollary 3.4.** *Let* r,s≥0r, *two non-negative integers and let* ξ∈Ξ. *Then the set*

$$\{x\in(\mathbf{B}_K^+)_{E,\xi}\mid q^r\varphi(x)\in q^s(\mathbf{B}_K^+)_{E,\xi}\}$$

*is equal to* $(\mathbf{B}_K^+)_{E,\xi}$ *if* $r\geq s$, *and is equal to* $\pi^{s-r}(\mathbf{B}_K^+)_{E,\xi}$ *if* $r<s$. □

**Corollary 3.5.** *For each integer* i≥0 *, the set*

$$\mathrm{Fil}^i N_{\underline{a}}=\{x\in N_{\underline{a}}\mid\varphi_{N,\underline{a}}(x)\in q^i N_{\underline{a}}\}$$

*is equal to the set of the elements* ${}^t((x_{1,\xi})_\xi,\ldots,(x_{n,\xi})_\xi)\in N_{\underline{a}}=(\mathbf{B}_K^+)_E^{\oplus n}$ n*such that* $x_{j,\sigma_\Xi^{-1}(\xi)}\in\pi^{i-r_{j,\xi}}(\mathbf{B}_K^+)_{E,\xi}$, *for every* (j,ξ)∈[1,n]×Ξ(j,ξ)∈[1,n]×Ξ*with* $\xi\neq\xi_0$ *and* $i>r_{j,\xi}$, *and that* $x_{w(j),\sigma_\Xi^{-1}(\xi_0)}\in\pi^{i-r_{j,\xi_0}}(\mathbf{B}_K^+)_{E,\xi_0}$. □

## 3.1. Semi-induced representations

**Definition 3.6.** We say that a free *E*-representation *V* is *semi-induced* if *V* is isomorphic to a direct sum $\oplus_i V_i$⊕iVi of free *E* -representations of $G_K$GK such that for each *i* , there exists a finite extension $E'_i$ of *E* such that $E'_i\otimes_E V_i$ is induced from an $E'_i$ valued continuous character of an open subgroup of $G_K$

Let *L* be a finite unramified extension of *K* and let $O_L$ denote its ring of integers. For ξ∈Ξ, the $O_E\otimes_{\mathbb{Z}_p}O_L$ -algebra $W_\xi\otimes_W O_L$ is equal to a direct product $W_\xi\otimes_W O_L=\prod_{\omega\in\Omega_\xi}W_{\xi,\omega}$, where for each $\omega\in\Omega_\xi$ω∈Ωξ, the $O_E$OE-algebra $W_{\xi,\omega}$ is the ring of integers of a finite unramified extension of $K_\xi$, and the index set $\Omega_\xi$Ωξ is a finite set whose cardinality is equal to the greatest common divisor of $[F:F_p]/c$[F:Fp]/c and [L:K][L:K].

**Proposition 3.7.** *Let V be a crystalline free E-representation V of* $G_K$*of dimension n. Let* $D_{crys}(V)_E$Dcrys(V)E*be the filtered φ-module over* $E\otimes_{Q_p}K$E⊗QpK*associated to V. Then V is semi-induced if there exist a basic datum* $(w, \underline{r}, \underline{\alpha}) = (w, (r_{i,\xi}), (\alpha_{i,\xi})) \in$ *,nof degree n and an* $E\otimes_{Q_p}K$ *-basis* $d_1,\ldots,d_n$d1,...,dn*of* $D_{crys}(V)_E$Dcrys(V)E*which satisfy the following properties:*

(1)*For each* $i\in Z$i∈Z*, the submodule* $Fil^iD_{crys}(V)_E$FiliDcrys(V)E*is equal to the set of elements* $\sum_{j=1}^{n} x_j d_j \in D_{crys}(V)_E$ E*such that the component of* $x_j$xj*at* $\sigma_{\Xi}^{-1}(\xi)$ *is equal to zero for every* (j,ξ)∈[1,n]×Ξ(j,ξ)∈[1,n]×Ξ*with* $\xi\neq\xi_0$ *and* $i>r_{j,\xi}$, ξ, *and that the component of* $x_{w(j)}$xw(j)*at* $\sigma_{\Xi}^{-1}(\xi_0)$ *is equal to zero for every* j∈[1,n] *with* $i>r_{j,\xi_0}$i>rj,ξ0.

(2)*For* ξ∈Ξ*, the component at ξ of the matrix of φ with respect to the basis* $d_1,\ldots,d_n$d1,...,dn*is equal to*

*for* $\xi\neq\xi_0$ξ≠ξ0*and is equal to*

$$\mathrm{diag}\left(\alpha_{1,\xi}p^{r_{1,\xi}},\ldots,\alpha_{n,\xi}p^{r_{n,\xi}}\right)\cdot C$$

*for* $\xi=\xi_0$ *where C denotes the permutation matrix corresponding to* $w^{-1}$w−1.

*Moreover the converse is true if V has distinct Hodge–Tate weights or if we replace V by* $E'\otimes_E V$ *for some finite extension* E′*of E.*

**Definition 3.8.** For a basic datum $(w, \underline{r}, \underline{\alpha}) \in \mathrm{Data}_{K,E,n}$ and for a crystalline free *E*-representation *V* of $G_K$, we say that a *V* is *semi-induced of type* $(w,\underline{r},\underline{\alpha})$ if *V* satisfies the conditions in Proposition 3.7 for $(w,\underline{r},\underline{\alpha})$.

**Remark 3.9.** We remark that, for a free $E$-representation $V$ of $G_K$GK which is crystalline and semi-induced, a basic data (w,r̲,α̲) as in Proposition 3.7 is not necessarily unique.

To prove Proposition 3.7 we need the following lemma.

**Lemma 3.10.** *A free E-representation V of* $G_K$GK*is semi-induced and crystalline if and only if V is isomorphic to a direct sum* $\oplus_i V_i$ i*of free E-representations such that for each i there exist a finite extension* Ei′*of E, a finite unramified extension* $K_i$ *of K, and a crystalline character* ηi:GKi→Ei′×*such that* Ei′⊗EVi*is isomorphic to the induced representation* $\mathrm{Ind}_{K_i}^{K}\eta_i$.

**Proof.** The "if" part of the claim is clear. We prove the "only if" part. Suppose that $V$ is semi-induced and crystalline. Then $V$ is isomorphic to a direct sum $\oplus_i V_i$⊕iVi of free $E$-representations such that for each $i$ , there exist a finite extension $E'_i$ of $E$ and a finite extensions $K_i$ of $K$ and a continuous character $\eta_i : G_{K_i} \to E_i'^{\times}$ such that $E'_i \otimes_E V_i$ is isomorphic to the induced representation $\mathrm{Ind}_{K_i}^{K}\eta_i$. Since $V$ is crystalline, $\mathrm{Ind}_{K_i}^{K}\eta_i$ is crystalline for every $i$ . This in particular shows that $\eta_i$ηi is crystalline. Since the inertia subgroup of $G_K$GK trivially acts on $D_{\mathrm{pst}}(\mathrm{Ind}_{K_i}^{K}\eta_i)_{E'_i}$ ′, the extension $K_i/K$ is unramified. This proves the claim. □

**Proof of Proposition 3.7.** Below we prove the "converse" part. The other part is easier to prove and follows by reversing the argument below.

Suppose that $V$ is crystalline and semi-induced. By Lemma 3.10, $V$ is isomorphic to a direct sum $\oplus_i V_i$⊕iVi of free $E$-representations such that for each $i$ there exist a finite extension $E'_i$ of $E$ and a finite unramified extensions $K_i$Ki and a crystalline character $\eta_i : G_{K_i} \to E_i'^{\times}$ such that $E'_i \otimes_E V_i$ is isomorphic to the induced representation $\mathrm{Ind}_{K_i}^{K}\eta_i$ . Let us fix $i$ . By enlarging $E'_i$ if necessary, we may assume that the residual degree of $E'_i/\mathbb{Q}_p$ is divisible by Let $\mathrm{rec} : K_i^{\times} \to G_{K_i}^{\mathrm{ab}}$ denote the reciprocity map in the local class field theory which sends a uniformizer in $K_i$Ki to a lift of the geometric Frobenius, and let

$I(K_i)$I(Ki) denote the set of embeddings $K_i \to E_i'$ of $\mathbb{Q}_p$-algebras. It follows from the classification of crystalline characters that for each $\iota \in I(K_i)$, there exists an integer $r_\iota \in \mathbb{Z}$r such that the character $\eta_i \circ \mathrm{rec}$ηi∘rec is equal to the tensor product of an unramified quasi-character $\kappa_i : K_i^\times \to E_i'^\times$ and the quasi-character which sends $x \in K_i^\times$ to the product

$$\prod_{\iota \in I(K_i)} \iota(x)^{r_\iota} \in E_i'^\times.$$

The tensor product $E_i' \otimes_{\mathbb{Q}_p} K_i$ is a product $E_i' \otimes_{\mathbb{Q}_p} K_i = \prod_{\iota \in I(K_i)} E_i'$ of copies of $E_i'$Ei′ where $x \otimes y \in E_i' \otimes_{\mathbb{Q}_p} K_i$ in the left hand side corresponds to $(\iota(x)y)_{\iota \in I(K_i)}$ in the right hand side. Let $\sigma_i$σi denote the Frobenius automorphism of $K_i$Ki. The composition with $\sigma_i^{-1}$ induces a cyclic permutation of the set $I(K_i)$I(Ki) which we denote by $\sigma_{I(K_i)}: I(K_i) \to I(K_i)$σ.

With the above notation, the filtered $\varphi$ -module $D_{\mathrm{crys}}(\eta_i)_{E_i'}$ ′ over $E_i' \otimes_{\mathbb{Q}_p} K_i$ has the following description. For $\iota \in I(K_i)$, let $D_{\mathrm{crys}}(\eta_i)_\iota$ denote the component at $\iota$ of $D_{\mathrm{crys}}(\eta_i)_{E_i'}$′. Then for $j \in \mathbb{Z}$ and for $\iota \in I(K_i)$, the component at $\iota$ of $\mathrm{Fil}^j D_{\mathrm{crys}}(\eta_i)_{E_i'}$ ′ is equal to $D_{\mathrm{crys}}(\eta_i)_\iota$ for $j \leq r$, and is equal to zero for $j > r_\iota$. For each $\iota \in I(K_i)$, take an $E_i'$ ′-basis of $D_{\mathrm{crys}}(\eta_i)_\iota$. Let $\iota \in I(K_i)$ and put $\iota' = \sigma_{I(K_i)}^{-1}(\iota)$ι′=σI(Ki)−1(ι). Let $b_\iota \in E_i'$bι∈Ei′ denote the unique element satisfying $\varphi(d_{\iota'}) = b_\iota d_\iota$φ(dι′)=bιdι. Then we have $\prod_{\iota \in I(K_i)} b_\iota = \kappa_i(p)$. Since the character $\eta_i$ηi takes values in $O_{E_i'}^\times$OEi′×, the element $p^{-\Sigma_\iota r_\iota}\kappa_i(p)$ is in $O_{E_i'}^\times$ Hence by changing the basis $d_\iota$dι, we may assume that $b_\iota \in p^{r_\iota} O_{E_i'}^\times$ for each $\iota \in I(K_i)$.

Then the claim follows from the observation that $D_{\mathrm{crys}}(\mathrm{Ind}_{K_i}^K \eta_i)_{E_i'}$ Dcrys(IndKiKηi)Ei′ is equal to $D_{\mathrm{crys}}(\eta_i)_{E_i'}$Dcrys(ηi)Ei′ regarded as a filtered $\varphi$ -module over $E_i' \otimes_{\mathbb{Q}_p} K$Ei′⊗QpK. □

**Definition 3.11.** For a basic datum $(w, \underline{r} = (r_{i,\xi}), \underline{\alpha} = (\alpha_{i,\xi})) \in \mathrm{Data}_{K,E,n}$ n and for a crystalline free $E$-representation $V$ of $G_K$GK, we say that $V$ is *nearly semi-induced of type* $(w, \underline{r}, \underline{\alpha})$ if the filtered $\varphi$ -module

$D_{crys}(V)_E$ over $E\otimes_{Qp}K$E⊗QpK associated to $V$ has a basis $d_1,\ldots,d_n$d1,...,dn over $E\otimes_{Qp}K$E⊗QpK satisfying the following conditions:

(1)For each i∈Z, the submodule $Fil^iD_{crys}(V)_E$FiliDcrys(V)E is equal to the set of elements $\sum_{j=1}^{n} x_j d_j \in D_{crys}(V)_E$ such that the component of $x_j$ at $\sigma_{\Xi}^{-1}(\xi)$ is equal to zero for every (j,ξ)∈[1,n]×Ξ with $\xi\neq\xi_0$ and $i>r_{j,\xi}$ and that the component of $x_{w(j)}$xw(j) at $\sigma_{\Xi}^{-1}(\xi_0)$ is equal to zero for every j∈[1,n] with $i>r_{j,\xi 0}$.

(2) Let *m* be the integer introduced in Section 2.4. (Note that we have m≤(r−1)/(p−1) Then for each ξ∈Ξ there exists a matrix $U_\xi\in GL_n(W_\xi)$ which is congruent to the identity matrix modulo $p^m m_E W_\xi$ such that the component at $\xi$ of the matrix of the endomorphism $\varphi$ of $D_{crys}(V)_E$Dcrys(V)E with respect to the basis $d_1,\ldots,d_n$ is equal to

$$U_\xi \cdot \mathrm{diag}\big(\alpha_{1,\xi}p^{r_{1,\xi}},\ldots,\alpha_{n,\xi}p^{r_{n,\xi}}\big)$$

for $\xi\neq\xi_0$ξ≠ξ0 and is equal to

$$U_{\xi_0} \cdot \mathrm{diag}\big(\alpha_{1,\xi}p^{r_{1,\xi}},\ldots,\alpha_{n,\xi}p^{r_{n,\xi}}\big)\cdot C$$

for $\xi=\xi_0$.

The following Proposition is an immediate consequence of Corollary 3.5.

**Proposition 3.12.** *Let* (w,r̲,α̲)∈DataK,E,n*be an effective basic datum of degree n and let V be a free E-representation of* $G_K$GK*which is crystalline and nearly semi-induced of type* (w,r̲,α̲). *Then there exists a tuple* a̲∈Tuple*such that the Wach module* **N**(V) *over* $(\mathbf{B}_K^+)_E$ *isomorphic to the Wach module* Na̲. □

The following Corollary is an immediate consequence of the Proposition 3.12 and Lemma 3.2.

**Corollary 3.13.** *Let V be as in Proposition 3.12. Then the semi-simplification of the reduction mod* $\mathfrak{m}_E$ *of any* $G_K$*-stable* $O_E$*-lattice of V is isomorphic to the semi-simplification of* $\mathbb{F} \otimes_{O_E} T_{\underline{\Omega}}$ *as a free* FF-*representation of* $G_K$.

**Remark 3.14.** The condition $U_\xi \equiv 1 \pmod{p^m \mathfrak{m}_E W_\xi}$ in Definition 3.11 implies that, for the representation $V$ in Proposition 3.12 and Corollary 3.13, the characteristic polynomial ch(Frob)ch(Frob) of the crystalline Frobenius Frob:=$\varphi^{[K:\mathbb{Q}_p]}$ on $D_{crys}(V)_E$ is $p$ -adically close to the characteristic polynomial ch(Qw,$\underline{r}$,$\underline{\alpha}$) of the matrix

$$Q_{w,\underline{r},\underline{\alpha}} := \operatorname{diag}\left(\prod_\xi \alpha_{1,\xi} p^{r_{1,\xi}}, \ldots, \prod_\xi \alpha_{n,\xi} p^{r_{n,\xi}}\right) w^{-1}.$$

It is remarkable that for some two dimensional crystalline representations of $G_{\mathbb{Q}_p}$GQp, for which ch(Frob)ch(Frob) are not $p$ -adically close to $\mathrm{ch}(Q_{w,\underline{r},\underline{\alpha}})$ for any $(w, \underline{r}, \underline{\alpha}) \in \mathrm{Data}_{\mathbb{Q}_p,E,2}$, the semi-simplification of its reduction mod $\mathfrak{m}_E$mE is explicitly obtained in [BB] and [BG] by a completely different method, based on the compatibility of the $p$-adic and the mod $p$ local Langlands correspondences. The latter method is effective for the other extreme case where ch(Frob)ch(Frob) are "nearly $p$ -adically farthest" from ch(Qw,$\underline{r}$,$\underline{\alpha}$) for any (w,$\underline{r}$,$\underline{\alpha}$). The limitations n=2 and K=$\mathbb{Q}_p$ in the latter method come from the fact that the $p$-adic local Langlands correspondence has been established only under these conditions. Hopefully they will be relaxed once the $p$-adic local Langlands correspondence will be established for other cases. However no promising method seems to be known [4] except for these two extreme cases.

# 4. LOCAL DEFORMATION RINGS AND MODULI OF WACH MODULES

In this section, we study universal family of Wach modules, local deformation rings, and subfamily corresponding to the family constructed in the previous section.

Let $AR_{O_E}$AROE denote the category of Artinian local $O_E$OE-algebra with residue field FF. Let $V_F$VF be a free FF-representation of $G_K$GK of dimension $n$ . We consider the functor $D_{VF}$DVF from $AR_{OE}$AROE the category of groupoids which sends an object $A$ in $AR_{OE}$AROE to the groupoid $D_{VF}(A)$DVF(A) of deformations of $V_F$VF to $A$ , that is, the groupoid $D_{VF}(A)$DVF(A) such that the objects in $D_{VF}$DVF are the pairs of a free $A$ -representation $V_A$VA of $G_K$GK and an isomorphism $V_A \otimes_A \mathbb{F} \xrightarrow{\cong} V_{\mathbb{F}}$ of free FF-representations of $G_K$GK, and for two objects $V_{A,1}$VA,1 and $V_{A,2}$VA,2 of $D_{VF}(A)$DVF(A), the morphisms from $V_{A,2}$ to $V_{A,2}$ are the isomorphisms $V_{A,1} \xrightarrow{\cong} V_{A,2}$ of free $A$ -representations of $G_K$GK which are compatible with the given isomorphisms $V_{A,i} \otimes_A F \cong V_F$ for i=1,2. Fix an FF-basis of $V_F$. We also consider the functor DVF□ from $AR_{OE}$ to the category of sets which sends an object $A$ in $AR_{OE}$ to the set DVF□(A) of framed deformations (or, "liftings" in the terminology of [CHT]) of $V_F$VF to $A$ , that is, the set DVF□(A) of isomorphism classes of triples of a free $A$ -representation $V_A$VA of $G_K$, an isomorphism $V_A \otimes_A F \cong V_F$ of free FF-representations of $G_K$, and a lifting to $V_A$VA of the fixed basis of $V_F$VF. Here two such triples $V_{A,1}$ and $V_{A,2}$ are called isomorphic if there exists an isomorphism $V_{A,1} \xrightarrow{\cong} V_{A,2}$ which transports the given basis of $V_{A,1}$to the given basis of $V_{A,2}$. The functor $D^{\square}_{V_{\mathbb{F}}}$ is pro-representable by a complete local Noetherian $O_E$OE-algebra $R^{\square}_{V_{\mathbb{F}}}$

Take an object $A$ in $AR_{OE}$, and let $V_A$ be an object of the groupoid $D_{VF}(A)$. Let $M_A := \mathbf{D}(V_A^*)$ be the étale $(\varphi,\Gamma_K)$ -module over $(\mathbf{A}_K)_A$ associated to the $A$ -linear dual $V_A^*$VA⍰ of $V_A$. For a tuple $\underline{r} = (r_\xi)$ of non-negative integers indexed by ξ∈Ξξ∈Ξ, let us consider the functor $W_{V_A,\leq \underline{r}}$,≤r from the category of $A$-algebras to the category of sets, which sends an $A$-algebra $B$ to the set of Wach lattices in

$M_A\otimes_A B$MA⊗AB with Hodge–Tate weights in the tuple $[0,\underline{r}]$[0,r] of intervals. (see [K2, §3] for the definition of Wach lattices in $M_A\otimes_A B$MA⊗AB). By [K2, Proposition (3.5)] the functor $W_{V_A,\le\underline{r}}$ has the following properties: The functor $W_{V_A,\le\underline{r}}$ is represented by a projective $A$ -scheme $\mathscr{W}_{V_A,\le\underline{r}}$. If $R$ is any complete local Noetherian ring with residue field $\mathbb{F}$F, and $V_R$VR a deformation of $V_{\mathbb{F}}$VF to $R$, then there exists a projective $R$ -scheme $\Theta_R:\mathscr{W}_{V_R,\le\underline{r}}\to\operatorname{Spec}R$ t

(1)If $A$ is in $\mathrm{AR}_{\mathcal{O}_E}$AROE and R→A is a homomorphism of local $\mathcal{O}_E$OE-algebras inducing the identity on residue fields, then there exists a canonical isomorphism $\mathscr{W}_{V_R,\le\underline{r}}\otimes_R A\xrightarrow{\cong}\mathscr{W}_{V_A,\le\underline{r}}$ where $V_A=V_R\otimes_R A$VA=VR⊗RA.

(2)The map $\Theta_R:\mathscr{W}_{V_R,\le\underline{r}}\to\operatorname{Spec}R$ becomes a closed embedding after inverting $p$. If $A$ is a finite $E$ -algebra, then a homomorphism R→AR→A of $\mathcal{O}_E$OE-algebras factors through the coordinate ring of the scheme theoretic image of $\Theta_R$ΘR if and only if $V_A=V_R\otimes_R A$VA=VR⊗RA is a crystalline representation with Hodge–Tate weights in $[0,\underline{r}]$[0,r].

For $R=R^{\square}_{V_{\mathbb{F}}}$ let $R^{\square,\le\underline{r}}_{V_{\mathbb{F}}}$RVF□,≤r denote the coordinate ring of the scheme theoretic image of $\Theta_{R^{\square}_{V_{\mathbb{F}}}}$ For a finite local $\mathbb{Q}_p$ -algebra $A$ and a free $A$ -representation $V_A$ of $G_K$ which is crystalline with non-negative Hodge–Tate weights, we put

$$P_{V_A}(T):=\det{}_{K\otimes_{\mathbb{Q}_p}A}\left(T-\varphi^{[K:\mathbb{Q}_p]}\middle|D_{\mathrm{crys}}(V_A^*)\right),$$

and let $c_{V_A,i}$cVA,i denote the coefficient of $T^i$ in $P_{V_A}(T)$ for i=0,…,n−1i=0,…,n−1. It follows from the argument in [K2, Corollary (3.7)] that the $\mathcal{O}_E$OE-algebra $R^{\square,\le\underline{r}}_{V_{\mathbb{F}}}$ ≤r has the following properties.

(1) $R^{\square,\le\underline{r}}_{V_{\mathbb{F}}}[1/p]$ is formally smooth over $E$.

(2)For i=0,…,n−1i=0,…,n−1, there is a unique $c_i\in R^{\square,\le\underline{r}}_{V_{\mathbb{F}}}[1/p]$ such that for any finite $E$-algebra $A$ , and any $\mathcal{O}_E$OE-algebra homomorphism $h : R^{\square}_{V_{\mathbb{F}}} \to A$, the representation $V_A$VA obtained from the universal representation over $R^{\square}_{V_{\mathbb{F}}}$ by specialization by *h* satisfies $c_{V_A,i}=h(c_i)$. Moreover, the element $c_i$ci is contained in the normalization of $\mathrm{Im}\{R^{\square,\leq \underline{r}}_{V_{\mathbb{F}}} \to R^{\square,\leq \underline{r}}_{V_{\mathbb{F}}}[1/p]\}$

**Lemma 4.1.** *For any nc-tuple* $\underline{r} = (r_{i,\xi})_{(i,\xi)\in[1,n]\times\Xi}$ *of integers and for every* ξ∈Ξ, *there exists a unique quotient* $R^{\square,\underline{r}}_{V_{\mathbb{F}}}$ *r̲of* $R^{\square}_{V_{\mathbb{F}}}$ *with the following properties:*

(1)$R^{\square,\underline{r}}_{V_{\mathbb{F}}}$ *r̲is p-torsion free, and if* $R^{\square,\underline{r}}_{V_{\mathbb{F}}} \neq 0$, *then* $R^{\square,\underline{r}}_{V_{\mathbb{F}}}[1/p]$ *formally smooth over E of pure dimension* $n^2+([K:Q_p]/c)\cdot\sum_{\xi\in\Xi} m_\xi$ *where for* ξ∈Ξ, *let* $m_\xi$mξ*denote the number of pairs* (i,j)∈[1,n]×[1,n] *satisfying* $r_{i,\xi}>r_{j,\xi}$ri,ξ>rj,ξ.

(2)*If* $E'$E′*is a finite extension of E,* $x:R_{V_F}\to E'$x:RVF→E′*a morphism of E-algebras, and* $V_x$Vx*the representation of* $G_K$GK*obtained by specializing the universal* $R_{V_F}$RVF*-representation by x, then x factors through* $R^{\square,\underline{r}}_{V_{\mathbb{F}}}$ *r̲if and only if* $V_x$Vx*is crystalline with Hodge–Tate weights* $(r_{1,z(\xi)},\ldots,r_{n,z(\xi)})_{\xi\in\Xi_{E'}}$ *where* $z:\Xi_{E'}\to\Xi=\Xi_E$ *denotes the map induced by the inclusion* $W_{O_E}\subset W_{O_{E'}}$.

**Proof.** This is a special case of [K3, Theorem (3.3.8)]. □

Fix an effective basic datum $(w,\underline{r},\underline{\alpha}) = (w,(\underline{r}_{i,\xi}),(\alpha_{i,\xi})) \in \mathrm{Data}_{K,E,n}$. We restrict ourselves to the case where $V_F$VF is the FF-linear dual of $\mathbb{F}\otimes_{O_E} T_{\underline{0}}$ where $T_{\underline{0}}$ is as in Lemma 3.2. We fix an FF-basis of $V_F$VF. Let $\underline{a} \in \mathrm{Tuple}$a̲∈Tuple. By Lemma 3.2, we have a canonical isomorphism $\mathbb{F}\otimes_{O_E} T^*_{\underline{a}} \cong V_{\mathbb{F}}$F⊗OETa̲⍰≅VF. Let *B* be an $O_E$OE-basis of the $O_E$OE-dual $T^*_{\underline{a}}$Ta̲ of $T_{\underline{a}}$Ta̲ which is compatible with the fixed basis of $V_F$VF under the above isomorphism $\mathbb{F}\otimes_{O_E} T^*_{\underline{a}} \cong V_{\mathbb{F}}$. The pair (Ta̲,B) gives a homomorphism $R^{\square,\underline{r}}_{V_{\mathbb{F}}} \to O_E$ of local $O_E$OE-algebras. Let $x_{\underline{a},B} : R^{\square,\underline{r}}_{V_{\mathbb{F}}}[1/p] \to E$ denote its base change to *E*.

**Proposition 4.2.** *The connected component of* $\mathrm{Spec}\, R^{\square,\underline{r}}_{V_{\mathbb{F}}}[1/p]$ *which contains the E-rational point* $x_{\underline{a},B}$*does not depend on the choice of the pair* $(\underline{a},B)$(a̲,B).

**Proof.** First we prove that the connected component does not depend on the choice of $B$ . Let $(\underline{a}, B_1)$and $(\underline{a}, B_2)$be two pairs with the common first component. Then there exists an element $g$ in the kernel of $GL_n(O_E)$→$GL_n(F)$ such that $B_2=gB_1$, that is, the matrix of the identity endomorphism of $T^*_{\underline{a}}$ with respect to the basis $B_2$B2 in the source and the basis $B_1$B1 in the target is equal to $g$. We fix a uniformizer $\varpi$ of $E$ and write $g=1_n+\varpi Y$g=1n+ϖY. Let $R:=O_E[[t]]$ be the ring of formal power series with coefficients in $O_E$OE. For an object $A$ in $AR(O_E)$AR(OE) and for a homomorphism f:R→A of local $O_E$OE-algebras, the pair $(A \otimes_{O_E} T_{\underline{a}}, (1+f(t)Y)B_1)$ defines an element in $D^{\square}_{V_{\mathbb{F}}}(A)$DVF□(A). The association $f \mapsto (A \otimes_{O_E} T_{\underline{a}}, (1+f(t)Y)B_1)$ f↦(A⨂OETa̲,(1+f(t)Y)B1) yields a homomorphism $\psi:R_{VF}\to R$ψ:RVF→R of complete local $O_E$OE-algebras. We claim that the homomorphism $\psi$ factors through the quotient $R^{\square,\underline{r}}_{V_{\mathbb{F}}}$ r̲ and hence induces a homomorphism $\overline{\psi} : R^{\square,\underline{r}}_{V_{\mathbb{F}}} \to R$ψ¯:. For a∈ma∈m, let $f_a:R\to E$fa:R→E denote the continuous homomorphism of $O_E$OE-algebras which sends $T$ to $a$. It follows from the construction of the homomorphism $\psi$ that the composite $f_a\circ\psi$fa∘ψ factors through the quotient $R^{\square,\underline{r}}_{V_{\mathbb{F}}}$ for any a∈ma∈m. Then the claim follows since Weierstrass preparation theorem implies that the homomorphism $R\to\prod_{a\in m}E$R→∏a∈mE given by $f_a$fa is injective. It is clear that the homomorphism $x_{\underline{a},B_1}$xa̲,B1 (resp. xa̲,B2) is equal to the composite

$$R^{\square,r}_{V_{\mathbb{F}}}[1/p] \xrightarrow{\overline{\psi}} R[1/p] \to E$$

where the second map is the continuous homomorphism which sends $t$ to 0 (resp. 1). Since $\operatorname{Spec} R[1/p]$ is irreducible, the two points xa̲,B1 and $x_{\underline{a},B_2}$xa̲,B2 lie in the same irreducible component of $\operatorname{Spec} R^{\square,\underline{r}}_{V_{\mathbb{F}}}[1/p]$SpecRVF□,r̲[1/p].

Let $R$ be a quotient of the ring $O_E[[\underline{X}]] = O_E[[X_{i,j,\theta}(i,j \in [1,n], \theta \in \Theta)]]$ such that $R$ is a finite $O_E$OE-algebra. Let $s_R : O_E[[\underline{X}]] \to R$ R denote the canonical surjection. Let $P(\underline{X})_R$ and $G_{\gamma,R}$ denote the matrices obtained by applying the homomorphism

$\mathrm{id} \otimes s_R : (\mathbf{A}_K^+)_{\mathcal{O}_E[[\underline{X}]]} = \mathcal{O}_E[[\underline{X}]] \otimes_{\mathbb{Z}_p} \mathbf{A}_K^+ \to R \otimes_{\mathbb{Z}_p} \mathbf{A}_K^+$ + to the entries of $P(\underline{X})$ and $G_\gamma G\gamma$ respectively. Then the pair $(P(\underline{X})R, (G\gamma, R)\gamma \in \Gamma)$ gives a structure of étale $(\varphi, \Gamma_K)(\varphi, \Gamma K)$-module on $(R \otimes_{\mathbb{Z}p} \mathbf{A}_K)^{\oplus n}$. Let $T_R$ denote the free $R$ -representation of $G_K GK$ associated to this étale $(\varphi, \Gamma_K)$ -module. Let TR⍰ denote the $R$ -linear dual of $T_R TR$. The collection $(T_R^*)_R$, where $R$ runs over the quotients of the ring $\mathcal{O}_E[[\underline{X}]]$ which is finite over $\mathcal{O}_E OE$, forms a projective system. We put $T^*_{\mathcal{O}_E[[\underline{X}]]} = \varprojlim_R T_R^*$ and take an $\mathcal{O}_E[[\underline{X}]]$ -basis $B_{\mathcal{O}_E[[\underline{X}]]}$ which lifts the given basis of $T_{\mathbb{F}}^* = V_{\mathbb{F}}$ The pair $(T^*_{\mathcal{O}_E[[\underline{X}]]}, B_{\mathcal{O}_E[[\underline{X}]]})$ gives a framed deformation of $V_{\mathbb{F}} VF$ to $\mathcal{O}_E[[\underline{X}]]$ We put $r_{\xi,\max}=\max_{i\in[1,n]} r_{i,\xi}$ r$\xi$,max=maxi$\in$[1,n]ri,$\xi$ for $\xi \in \Xi$ and $\underline{r}_{\max} = (r_{\xi,\max})_{\xi \in \Xi}$. It follows from the construction of the $\mathcal{O}_E[[\underline{X}]]$ scheme $\mathscr{W} = \mathscr{W}_{T^*_{\mathcal{O}_E[[\underline{X}]]}, \leq \underline{r}_{\max}}$ ,≤rmax that the matrices $P(\underline{X})$ and $G_\gamma G\gamma$ give an $\mathcal{O}_E[[\underline{X}]]$-rational point of WW. Hence the homomorphism $R^{\square}_{V_{\mathbb{F}}} \to \mathcal{O}_E[[\underline{X}]]$ which comes from the universality factors through the quotient $R^{\square, \leq \underline{r}_{\max}}_{V_{\mathbb{F}}}$. Since the set of $E$-rational points of a rigid analytic open unit disc over $E$ is not contained in any proper analytic closed subvariety, a technique similar to that in the proof of the claim in the previous paragraph shows that the homomorphism $R_{V_{\mathbb{F}}} \to \mathcal{O}_E[[\underline{X}]]$ factors through the quotient $R^{\square, \underline{r}}_{V_{\mathbb{F}}}$. Since the scheme $\operatorname{Spec} \mathcal{O}_E[[\underline{X}]][1/p]$ is irreducible, the claim follows. □

# 5. APPLICATIONS TO AUTOMORPHY LIFTING THEOREM AND POTENTIAL AUTOMORPHY THEOREM

**Definition 5.1.**

We say that a crystalline free $E$-representation $V$ of $G_K GK$ is *absolutely nearly semi-induced* if there exist a finite extension $E'E'$ of $E$ and a basic datum $(w, \underline{r}, \underline{\alpha}) = (w, (r_{i,\xi}), (\alpha_{i,\xi})) \in \mathrm{Data}_{K,E',n}$ such that $E' \otimes_E V$ is nearly semi-induced of type $(w, \underline{r}, \underline{\alpha})$. We say that a free

*E*-representation *V* of $G_K$ is *upper-triangular* if there is a filtration on *V* by *E*-subrepresentations whose graded pieces are one-dimensional.

Proposition 4.2 together with Lemma 1.4.1 and Lemma 1.4.3 of [BLGGT] gives the following result.

**Proposition 5.2.** *Let V be a free E-representation of* $G_K$ *which is either crystalline and absolutely nearly semi-induced or potentially crystalline and upper-triangular. Then* $V \otimes_E O_{\overline{\mathbb{Q}}_p}$ *is potentially diagonalizable in the sense of [BLGGT, 1.4].* □

This proposition, together with the main results of [BLGGT, Theorem 4.2.1, Corollary 4.5.2], we have the following automorphy lifting and potential automorphy results.

**Corollary 5.3.** *Let F be an imaginary CM field and let* $F^+$ *denote its maximal totally real subfield. Let* $n \geq 1$ *be an integer and let p be a prime number such that F is unramified at p, and that F does not contain a primitive p-th root of unity. Fix an isomorphism* $\overline{\mathbb{Q}}_p \xrightarrow{\cong} \mathbb{C}$. $\iota$*. Let E be a finite extension of* $\mathbb{Q}_p$ *and let* $(r,\mu)$ *be the pair of a continuous absolutely irreducible representation* $r: G_F \to GL_n(E)$ *and a continuous character* $\mu: G_{F^+} \to E^\times$ *that* $r^c$ *is isomorphic to* $r^\vee \otimes \mu$ *where* $c \in Gal(F/F^+)$ *denotes the complex conjugation. Let* $\overline{r}$ *denote the reduction of r and assume that the following conditions are satisfied:*

(1) *r is unramified outside a finite set of primes.*

(2) *The restriction of* $\overline{r}$ *to* $G_{F(\zeta_p)}$ *is irreducible.*

(3) $p \geq 2(d+1)$ *where d is the maximal dimension of an irreducible subrepresentation of the restriction of* $\overline{r}$ *to the closed subgroup of* $G_F$ *generated by all Sylow pro-p subgroups.*

(4) *For all prime v of F dividing p, the restriction* $r|_{G_{F_v}}$ *is potentially crystalline with distinct Hodge–Tate weights and is either crystalline and absolutely nearly semi-induced or upper-triangular.*

(5)*There is a RAECSDC automorphic representation* $(\pi,\nu)$ *of* $GL_n(A_F)$ *whose associated pair* $(r_{p,\iota}(\pi),r_{p,\iota}(\nu))$ *of a p-adic Galois representation* $r_{p,\iota}(\pi)$ *and a p-adic character* $r_{p,\iota}(\nu)$ *satisfies the following conditions:*

- $(\bar{r},\bar{\mu})$*is isomorphic to* $(\bar{r}p,\iota(\pi),\bar{r}p,\iota(\nu))$.
- *either π is ι-ordinary in the sense of [G, Definition 5.1.2.], or for every* v|pv|p, *the restriction* $r_{p,\iota}(\pi)|_{GFv}$rp,ι(π)|GFv*is either crystalline and absolutely nearly semi-induced or potentially crystalline and upper-triangular.*

*Then* (r,μ)(r,μ)*is automorphic of level potentially prime to p.*

**Corollary 5.4.** *Let* $F^+$F+*be a totally real field. Let* n≥1n≥1*be an integer. Let p be a prime number such that* p≥2(n+1)p≥2(n+1)*and p is unramified in* $F^+$F+. *Let E be a finite extension of* $Q_p$Qp*and let* (r,μ)(r,μ)*be the pair of a continuous absolutely irreducible representation* $r:G_{F^+}\to GL_n(E)$r:GF+→GLn(E)*and a continuous character* $\mu:G_{F^+}\to E^\times$μ:GF+→E×*such that the pair* (r,μ)(r,μ)*is a totally odd, essentially conjugate self dual representation. Let* $\bar{r}$r̄*denote the reduction of r and assume that the following conditions are satisfied:*

(1)*r is unramified outside a finite set of primes.*

(2)*The restriction of* $\bar{r}$r̄*to* $G_{F^+(\zeta_p)}$GF+(ζp)*is irreducible.*

(3)*For all prime v of* $F^+$F+*dividing p, the restriction* r|GFv+*is potentially crystalline with distinct Hodge–Tate weights and is either crystalline and absolutely nearly semi-induced or upper-triangular.*

*Then there exists a Galois, totally real extension* $F^{+'}/F^+$F+′/F+*such that restriction* $(r|_{GF^{+'}},\mu|_{GF^{+'}})$(r|GF+′,μ|GF+′)*is automorphic of level prime to p.*

## ACKNOWLEDGMENTS

This paper is a generalized version of the preprint (2008) of the first author. He thanks the second author for discussions and for extending the construction of the family of Wach modules in the previous version, and for checking and correcting the other part. (He also thanks him for precisely citing the recent preprints to remove some parts of the previous version. Especially, by citing [BLGGT], we could remove the part showing $R^{red}$=TRred=T for $GL_n$GLn case.) He thanks Ivan Fesenko in Nottingham University for the hospitality from April/2008 to March/2010.

He sincerely thanks TOYOTA Central R&D Labs., Inc. for offering him a special position in which he can concentrate on pure mathematical research. The president of the company at that time told him that if they forced him to do something else, then it would be against the policy of the company, and it would be against the philosophy of the founder of the company as well. He also sincerely thanks Sakichi Toyoda for his philosophy, and the executives for inheriting it from him for 80 years after his death.

## REFERENCES

1. [BLGGT] T. Barnet-Lamb, T. Gee, D. Geraghty, R. Taylor, Potential automorphy and change of weight, Ann. of Math. (2) 179 (2014) 501–609.

2. [BLGHT] T. Barnet-Lamb, D. Geraghty, M. Harris, R. Taylor, A family of Calabi–Yau varieties and potential automorphy II, Publ. Res. Inst. Math. Sci. 47 (1) (2011) 29–98.

3. [B1] L. Berger, Limites de représentations cristallines, Compos. Math. 140 (2004) 1473–1498.

4. [B2] L. Berger, Représentations modulaires de GL2(Qp) et représentations galoisiennes de dimension 2, Astérisque 330 (2010) 263–279.

5. [B3] L. Berger, Local constancy for the reduction mod p of 2-dimensional crystalline representations, Bull. Lond. Math. Soc. 44 (3) (2012) 451–459.

6. [BB] L. Berger, C. Breuil, Sur la réduction des représentations cristallines de dimension 2 en poid moyens, unpublished note, which is contained in [B2].

7. [BLZ] L. Berger, H. Li, H.J. Zhu, Construction of some families of 2-dimensional crystalline representations, Math. Ann. 329 (2) (2004) 365–377.

8. [BG] K. Buzzard, T. Gee, Explicit reduction modulo p of certain two-dimensional crystalline representations, Int. Math. Res. Not. IMRN (12) (2009) 2303–2317.

9. [CHT] L. Clozel, M. Harris, R. Taylor, Automorphy for some -adic lifts of automorphic mod Galois representations, Publ. Math. Inst. Hautes Etudes Sci. 108 (2008) 1–181.

10. [D] G. Dousmanis, On reductions of families of crystalline Galois representations, Doc. Math. 15 (2010) 873–938.

11. [F] J.-M. Fontaine, Représentations p-adiques des corps locaux I. The Grothendieck Festschrift, vol. II, Progr. Math., vol. 87, Birkhäuser Boston, Boston, MA, 1990, pp. 249–309.

12. [GL] H. Gao, T. Liu, A note on potential diagonalizability of crystalline representations, preprint, http://arxiv.org/abs/1204.4704.

13. [G] D. Geraghty, Modularity lifting theorems for ordinary Galois representations, PhD thesis, Harvard University, ISBN 978-1124-07900-4, 2010, 131 pp., ProQuest LLC.

14. [HSBT] M. Harris, N. Shepherd-Barron, R. Taylor, A family of Calabi–Yau varieties and potential automorphy, Ann. of Math. 171 (2010) 779–813.

15. [K2] M. Kisin, Modularity for some geometric Galois representations, in: L-Functions and Galois Representations, Durham, 2004, pp. 438–470.

16. [K3] M. Kisin, Potentially semistable deformation rings, J. Amer. Math. Soc. 103 (2008) 513–546.

17. [K1] M. Kisin, Moduli of finite flat group schemes and modularity, Ann. of Math. 170 (2009) 1085–1180.

18. [T] R. Taylor, Automorphy for some -adic lifts of automorphic mod Galois representations II, Publ. Math. Inst. Hautes Etudes Sci. 108 (2008) 183–239.

19. [YY] G. Yamashita, S. Yasuda, Reduction of two dimensional crystalline representations and Hypergeometric polynomials, in preparation.

20. [Z] H.J. Zhu, Crystalline representations of GQpa with coefficients, preprint, http://arxiv.org/ abs/0807.1078.

5

# SOME COMMON FIXED POINT THEOREMS SATISFYING ($\Psi - \Phi$) MAPS IN PARTIAL METRIC SPACES

**Reza Arab**

Department of Mathematics, Sari Branch, Islamic Azad University, Sari, Iran

## ABSTRACT

In this paper, we present some coincidence and common fixed point results for infinite families of contractive maps satisfying a new class of pairs of generalized contractive type mappings defined in partial metric spaces. Our results extend and generalize many known results in the literature. Also, we introduce an example to support the validity of our results.

## KEYWORDS

partial metric, common fixed point, coincidence point, generalized contraction

# 1. INTRODUCTION AND PRELIMINARIES

In 1994, in [8] Matthews introduced the notion of a partial metric space and proved the contraction principle of Banach in this new framework. The Existence of a fixed point for the contraction type mappings in partially metric spaces and its applications have been considered recently by many authors [1,2,3,6,7,9,12,13,15]. Consistent with [4, 8], the following de_nitions and results will be needed in the sequel.

**Definition 1.1.** [8] *A partial metric on a nonempty set* $X$ *is a function* $p: X \times X \to R^+$ *such that for all* $x, y, z \in X$ *:*

($P_1$) $x = y \Leftrightarrow p(x, x) = p(x, y) = p(y, y)$,

($P_2$) $p(x, x) \leq p(x, y)$,

($P_3$) $p(x, y) = p(y, x)$,

($P_4$) $p(x, y) \leq p(x, z) + p(z, y) - p(z, z)$.

*A partial metric space is a pair (X, p) such that X is a nonempty set and p is a partial metric on X.*

It is clear that, if $p(x, y) = 0$, then, we get $x = y$ from ($P_1$) and ($P_2$). But if $x = y$, $p(x, y)$ may be a positive number other than 0.

**Example 1.2.** [8] Let a function $p : R^+ \times R^+ \to R^+$ be defined by $p(x, y) = \max\{x, y\}$ for any $x, y \in R^+$. Then, $(R^+, p)$ is a partial metric space.

**Example 1.3.** [8] *If* $X = \{[a, b] : a, b$ ⍰ $R, a \leq b\}$, *then* $p : X \times X \to R^+$ *de_ned by* $p([a, b], [c, d]) = \max\{b, d\} - \min\{a, c\}$ *is a partial metric on X.*

If $p$ is a partial metric on $X$, then the function $p^s : X \times X \to R^+$ given by

$$p^S(x,y) = 2p(x,y) - p(x,x) - p(y,y) \tag{1.1}$$

is a metric on *X*.

**Definition 1.4.** [8, 10, 11] *Let (X, p) be a partial metric space. Then*

(i) A sequence {xn} in a partial metric space (X, p) converges to a point x ∈ X if and only if $p(x,x) = \lim_{n\to\infty} p(x,x_n)$.

(ii) A sequence {xn} in a partial metric space (X, p) is called a Cauchy sequence if there exists (and is finite $\lim_{n,m\to\infty} p(x_m,x_n)$.

(iii) A partial metric space (X, p) is said to be complete if every Cauchy sequence {xn} in X converges to a point x ∈ X, that is $p(x,x) = \lim_{n,m\to\infty} p(x_m,x_n)$.

It is easy to see that, every closed subset of a complete partial metric space is complete.

**Lemma 1.5.** [8, 9, 10] *Let (X, p) be a partial metric space. Then*

*(a) $\{x_n\}$ is a Cauchy sequence in (X, p) if and only if it is a Cauchy sequence in the metric space (X, $p^s$).*

*(b) A partial metric space (X, p) is complete if and only if the metric space (X, $p^s$) is complete. Further-more,*

$$\lim_{n\to\infty} p^S(x_n,x) = 0$$

*if and only if*

$$p(x,x) = \lim_{n\to\infty} p(x,x_n) = \lim_{n,m\to\infty} p(x_m,x_n).$$

**Lemma 1.6.** [2] A mapping f : X −→ X is said to be continuous at a ∈ X, if for every ϵ > 0, there existsδ > 0 such that f(B(a, δ)) ⊂ B(f(a), ϵ).

The following result is easy to check.

**Lemma 1.7.** Let (X, p) be a partial metric space. $T : X \to X$ is continuous if and only if given a sequence $\{x_n\} \subseteq X$ and $x \in X$ such that $p(x,x) = \lim_{n\to\infty} p(x_n, x)$, then $p(Tx,Tx) = \lim_{n\to\infty} p(Tx_n, Tx)$.

**Definition 1.8.** Let X be a set, T and g are selfmaps of X. A point x in X is called a coincidence point of T and g if Tx = gx. We shall call w = Tx = gx a point of coincidence of T and g.

**Definition 1.9.** [14] Let (X, p) be a partial metric space and $T, g : X \to X$ are mappings of X into itself. We say that the pair {T, g} is partial compatible if the following conditions is held:

($b_1$) if p(x, x) = 0 then p(gx, gx) = 0,

($b_2$) $\lim_{n\to\infty} p(T(gx_n), g(Tx_n)) = 0$ whenever $\{x_n\}$ is a sequence in X such that $Txn \to t$ and $gxn \to t$ for some $t \in X$.

**Example 1.10.** Let X = $[0,\infty)$ endowed with the usual partial metric p de_ned by $p : X \times X \to X$ with p(x, y) =max{x, y}. Suppose T, $g : X \to X$ such that $Tx = 5x^3$ and $gx = 2x^3$ for all $x \in X$. It is easy to check that the pair {T, g} is partial compatible.

Matthews [8] obtained the following Banach fixed-point theorem on complete partial metric spaces.

**Theorem 1.11.** (Matthews[8]) Let f be a mapping of a complete dualistic partial metric space (X, p) into itself such that there is a real number c with $0 \le c < 1$, satisfying

$$p(f(x), f(y)) \le cp(x, y), \tag{1.2}$$

for all $x, y \in X$, then f has a unique fixed point.

Recently, I. Altun and A. Erduran [5] obtained the following nice result, which generalizes Theorem 1.11 of Matthews.

**Theorem 1.12.** [5] *Let (X, p) be a complete partial metric space and let $f : X \to X$ be a map such that*

$$p(f(x), f(y)) \leq \phi p(x, y), \tag{1.3}$$

for all x, $y \in X$, where $\phi : [0,\infty) \to [0,\infty)$ is a continuous, non-decreasing function such that $\phi(t) < t$ for each $t > 0$. Then f has a unique fixed point.

**Theorem 1.13.** [3] Let (X, p) be a complete partial metric space and let $F : X \to X$ be a map such that

$$p(Fx, Fy) \leq \phi\left( \max\left\{ \begin{array}{l} p(x,y), p(x,Fx), p(y,Fy), \\ \frac{1}{2}[p(x,Fy) + p(y,Fx)] \end{array} \right\} \right) \tag{1.4}$$

for all x, $y \in X$, where $\phi : [0,\infty) \dashrightarrow [0,\infty)$ is continuous, nondecreasing function such that $\phi(t) < t$ for each $t > 0$, then F has a unique fixed point.

## 2. MAIN RESULTS

In this section, we fixed the set of functions by $\psi, \varphi : [0,+\infty) \to [0,+\infty)$ such that

(a) $\psi$ is increasing;

(b) $\varphi(t) < \psi(t)$ for each $t > 0$, $\varphi(0) = \psi(0) = 0$;

(c) $\varphi(t)$ and $\psi(t)$ are continuous functions.

Define $\Psi = \{(\psi, \varphi) : \psi$ and $\varphi$ satisfy (a),(b) and (c)$\}$.

Now, we establish an existence of common fixed point of a family mappings satisfying contractive condition involving $\psi - \varphi$ functions in the setup of partial metric spaces. Our results generalize Theorems 1.11,1.12 and 1.13.

**Theorem 2.1.** Let (X, p) be a complete partial metric space. Suppose $\{T_\alpha : X \dashrightarrow X : \alpha \in I\}$ be a family of mappings and $g : X \to X$ be a self map. Also there exists $i_0 \in I$ such that $T_{i_0} \subseteq g(X)$, gX is closed in (X, p) and assume that there exists $(\psi, \varphi) \in \Psi$ such that

$$\psi\left(p\left(T_{i_0}x, T_i y\right)\right) \le \varphi\left(M(x,y)\right), \tag{2.1}$$

*for all x, y* ⍰ *X, where*

$$M(x,y) = \max\left\{\begin{matrix} p(gx,gy), p\left(gx, T_{i_0}x\right), p(gy, T_i y), \\ \dfrac{p(gx, T_i y) + p\left(gy, T_{i_0}x\right)}{2} \end{matrix}\right\}. \tag{2.2}$$

Then, there exists a unique $x \in X$ such that $gx = T_i x$ for all $i \in I$, that is, g and $\{T_i : i \in I\}$ have a unique coincidence point in X. Moreover, any coincidence point of g and $T_{i_0}$ is a coincidence point of g and $\{T_i : i \in I\}$. Also, we have $p(gx, gx) = 0$.

**Proof**. Let $x_0$ be an arbitrary point in *X*. Construct the sequence $\{x_n\}$ such that $T_{i_0}(x_n) = g(x_n+1)$ for each *n*= 0, 1, 2, .... which is possible since $T_{i_0}X \subseteq gX$. Now by (2.1), we have

$$\psi\left(p\left(T_{i_0}(x_n), T_{i_0}(x_{n+1})\right)\right) \le \varphi\left(M(x_n, x_{n+1})\right), \tag{2.3}$$

Where

$$M(x_n, x_{n+1})$$

$$= \max\left\{\begin{array}{l} p(g(x_n), g(x_{n+1})), p(g(x_n), T_{a0}(x_n)), \\ p(g(x_{n+1}), T_{i_0}(x_{n+1})), \\ \dfrac{p(g(x_n), T_{i_0}(x_{n+1})) + p(g(x_{n+1}), T_{i_0}(x_n))}{2} \end{array}\right\}$$

$$= \max\left\{\begin{array}{l} p(T_{i_0}(x_{n-1}), T_{i_0}(x_n)), p(T_{i_0}(x_{n-1}), T_{i_0}(x_n)), \\ p(T_{i_0}(x_n), T_{i_0}(x_{n+1})), \\ \dfrac{p(T_{i_0}(x_{n-1}), T_{i_0}(x_{n+1})) + p(T_{i_0}(x_n), T_{i_0}(x_n))}{2} \end{array}\right\}$$

$$= \max\left\{\begin{array}{l} p(T_{i_0}(x_{n-1}), T_{i_0}(x_n)), p(T_{i_0}(x_n), T_{i_0}(x_{n+1})), \\ \dfrac{p(T_{i_0}(x_{n-1}), T_{i_0}(x_{n+1})) + p(T_{i_0}(x_n), T_{i_0}(x_n))}{2} \end{array}\right\}.$$

By $(p_4)$, we have

$$\frac{p(T_{i_0}(x_{n-1}), T_{i_0}(x_{n+1})) + p(T_{i_0}(x_n), T_{i_0}(x_n))}{2}$$
$$\leq \frac{p(T_{i_0}(x_{n-1}), T_{i_0}(x_n)) + p(T_{i_0}(x_n), T_{i_0}(x_{n+1}))}{2}$$
$$\leq \max\{p(T_{i_0}(x_{n-1}), T_{i_0}(x_n)), p(T_{i_0}(x_n), T_{i_0}(x_{n+1}))\}$$

Therefore,

$$M(x_n, x_{n+1}) = \max\left\{\begin{array}{l} p(T_{i_0}(x_{n-1}), T_{i_0}(x_n)), \\ p(T_{i_0}(x_n), T_{i_0}(x_{n+1})) \end{array}\right\}.$$

If we suppose that

$$M(x_n, x_{n+1}) = p(T_{i_0}(x_n), T_{i_0}(x_{n+1}))$$

and $T_{i_0}(x_n) = T_{i_0}(x_{n+1})$ for some $n$, then by $T_{i_0}(x_n) = g(x_{n+1})$, $T_{i_0}(x_{n+1}) = g(x_{n+1})$, that is, $T_{i_0}$ and $g$ have a coincidence at $x = x_{n+1}$, and so the proof is completed, otherwise if $p(T_{i_0}(x_n), T_{i_0}(x_{n+1})) > 0$, then

$$\psi\left(p\left(T_{i_0}(x_n),T_{i_0}(x_{n+1})\right)\right)\le\varphi\left(p\left(T_{i_0}(x_n),T_{i_0}(x_{n+1})\right)\right)$$
$$<\psi\left(p\left(T_{i_0}(x_n),T_{i_0}(x_{n+1})\right)\right).$$

It is impossible.

If $M(x_n,x_{n+1})=p\left(T_{i_0}(x_{n-1}),T_{i_0}(x_n)\right)$, we get

$$\psi\left(p\left(T_{i_0}(x_n),T_{i_0}(x_{n+1})\right)\right)\le\varphi\left(p\left(T_{i_0}(x_{n-1}),T_{i_0}(x_n)\right)\right)$$
$$<\psi\left(p\left(T_{i_0}(x_{n-1}),T_{i_0}(x_n)\right)\right). \tag{2.4}$$

If $T_{i_0}(x_{n-1})=T_{i_0}(x_n)$ for some $n$, then by $T_{i_0}(x_{n-1})=g(x_n)$, $T_{i_0}(x_n)=g(x_n)$, that is, $T_{i_0}$ and $g$ have a coincidence at $x = x_n$, and so the proof is completed. Since we suppose that $p\left(T_{i_0}(x_{n-1}),T_{i_0}(x_n)\right)>0$ and by the monotone property of $\psi$ function, we have $p\left(T_{i_0}(x_n),T_{i_0}(x_{n+1})\right)<p\left(T_{i_0}(x_{n-1}),T_{i_0}(x_n)\right)$. Therefore, the sequence $p\left(T_{i_0}(x_n),T_{i_0}(x_{n+1})\right)$ of real numbers is monotone decreasing. Hence there exists a real number $r \geq 0$ such that,

$$\lim_{n\to\infty} p\left(T_{i_0}(x_n),T_{i_0}(x_{n+1})\right)=r.$$

We claim that $r = 0$. On the contrary, assume that

$$\lim_{n\to\infty} p\left(T_{i_0}(x_n),T_{i_0}(x_{n+1})\right)=r>0. \tag{2.5}$$

Since $\psi$ and $\varphi$ are continuous then from (2.4) and (2.5), we have

$$\psi(r)=\lim_{n\to\infty}\psi\left(p\left(T_{i_0}(x_n),T_{i_0}(x_{n+1})\right)\right)$$
$$=\lim_{n\to\infty}\varphi\left(p\left(T_{i_0}(x_{n-1}),T_{i_0}(x_n)\right)\right)=\varphi(r),$$

and so $r = 0$, a contradiction. Thus

$$\lim_{n\to\infty} p\left(T_{i_0}(x_n), T_{i_0}(x_{n+1})\right) = 0. \tag{2.6}$$

From $p(T_{i_0}(x_n), T_{i_0}(x_n))$, $p(T_{i_0}(x_{n+1}), T_{i_0}(x_{n+1})) \le p(T_{i_0}(x_n), T_{i_0}(x_{n+1}))$ and (2.6), we have

$$\lim_{n\to\infty} p\left(T_{i_0}(x_n), T_{i_0}(x_n)\right)$$
$$= \lim_{n\to\infty} p\left(T_{i_0}(x_{n+1}), T_{i_0}(x_{n+1})\right) = 0. \tag{2.7}$$

From (2.6), (2.7) and (1.1), we have

$$\lim_{n\to\infty} p^s\left(T_{i_0}(x_n), T_{i_0}(x_{n+1})\right) = 0. \tag{2.8}$$

Next, we claim that $\{T_{i_0} x_n\}$ is a Cauchy sequence in the metric space $(X, p^s)$. Assume the contrary. Then there exists an $\epsilon > 0$ and subsequences $\{x_{n(k)}\}$ and $\{x_{m(k)}\}$ of $\{x_n\}$ with $n(k) > m(k) > k$ such that $p(T_{i_0}(x_{n(k)}), T_{i_0}(x_{m(k)})) \ge \epsilon$ and $p(T_{i_0}(x_{n(k)-1}), T_{i_0}(x_{m(k)})) < \epsilon$.

Then we have

$$\begin{aligned}
\epsilon &\le p\left(T_{i_0}\left(x_{n(k)}\right), T_{i_0}\left(x_{m(k)}\right)\right) \\
&\le p\left(T_{i_0}\left(x_{n(k)}\right), T_{i_0}\left(x_{n(k)-1}\right)\right) \\
&\quad + p\left(T_{i_0}\left(x_{n(k)-1}\right), T_{i_0}\left(x_{m(k)}\right)\right) \\
&\quad - p\left(T_{i_0}\left(x_{n(k)-1}\right), T_{i_0}\left(x_{n(k)-1}\right)\right) \\
&< p\left(T_{i_0}\left(x_{n(k)}\right), T_{i_0}\left(x_{n(k)-1}\right)\right) + \epsilon \\
&\quad - p\left(T_{i_0}\left(x_{n(k)-1}\right), T_{i_0}\left(x_{n(k)-1}\right)\right),
\end{aligned} \tag{2.9}$$

Taking $k \to \infty$ in (2.9) and using (2.6) and (2.7) we get

$$\lim_{k\to\infty} p\left(T_{i_0}\left(x_{n(k)}\right), T_{i_0}\left(x_{m(k)}\right)\right)$$
$$= \lim_{k\to\infty} p\left(T_{i_0}\left(x_{n(k)-1}\right), T_{i_0}\left(x_{m(k)}\right)\right) = \epsilon. \tag{2.10}$$

Thus from the definition of $p$ we have

$$\begin{aligned}
&p\left(T_{i_0}\left(x_{m(k)}\right), T_{i_0}\left(x_{n(k)}\right)\right)\\
&\le p\left(T_{i_0}\left(x_{m(k)}\right), T_{i_0}\left(x_{m(k)-1}\right)\right)\\
&+p\left(T_{i_0}\left(x_{m(k)-1}\right), T_{i_0}\left(x_{n(k)-1}\right)\right)\\
&+p\left(T_{i_0}\left(x_{n(k)-1}\right), T_{i_0}\left(x_{n(k)}\right)\right)\\
&-p\left(T_{i_0}\left(x_{m(k)-1}\right), T_{i_0}\left(x_{m(k)-1}\right)\right)\\
&-p\left(T_{i_0}\left(x_{n(k)-1}\right), T_{i_0}\left(x_{n(k)-1}\right)\right),
\end{aligned} \tag{2.11}$$

$$\begin{aligned}
&p\left(T_{i_0}\left(x_{m(k)-1}\right), T_{i_0}\left(x_{n(k)-1}\right)\right)\\
&\le p\left(T_{i_0}\left(x_{m(k)-1}\right), T_{i_0}\left(x_{m(k)}\right)\right)\\
&+p\left(T_{i_0}\left(x_{m(k)}\right), T_{i_0}\left(x_{n(k)}\right)\right)\\
&+p\left(T_{i_0}\left(x_{n(k)}\right), T_{i_0}\left(x_{n(k)-1}\right)\right)\\
&= p\left(T_{i_0}\left(x_{m(k)}\right), T_{i_0}\left(x_{m(k)}\right)\right)\\
&-p\left(T_{i_0}\left(x_{n(k)}\right), T_{i_0}\left(x_{n(k)}\right)\right).
\end{aligned} \tag{2.12}$$

Taking $k \to \infty$ in (2.11) and (2.12) and using (2.6), (2.7) and (2.10) we get

$$\lim_{k\to\infty} p\left(T_{i_0}\left(x_{n(k)}\right), T_{i_0}\left(x_{m(k)}\right)\right)$$
$$= \lim_{k\to\infty} p\left(T_{i_0}\left(x_{n(k)-1}\right), T_{i_0}\left(x_{m(k)-1}\right)\right) = \epsilon. \tag{2.13}$$

Also,

$$\begin{aligned}\epsilon &\le p\left(T_{i_0}\left(x_{n(k)}\right), T_{i_0}\left(x_{m(k)}\right)\right)\\
&\le p\left(T_{i_0}\left(x_{n(k)}\right), T_{i_0}\left(x_{m(k)-1}\right)\right)\\
&+p\left(T_{i_0}\left(x_{m(k)-1}\right), T_{i_0}\left(x_{m(k)}\right)\right)\\
&-p\left(T_{i_0}\left(x_{m(k)-1}\right), T_{i_0}\left(x_{m(k)-1}\right)\right)\\
&\le p\left(T_{i_0}\left(x_{n(k)}\right), T_{i_0}\left(x_{n(k)-1}\right)\right)\\
&+p\left(T_{i_0}\left(x_{n(k)-1}\right), T_{i_0}\left(x_{m(k)-1}\right)\right)\\
&+p\left(T_{i_0}\left(x_{m(k)-1}\right), T_{i_0}\left(x_{m(k)}\right)\right)\\
&-p\left(T_{i_0}\left(x_{n(k)-1}\right), T_{i_0}\left(x_{n(k)-1}\right)\right)\\
&-p\left(T_{i_0}\left(x_{m(k)-1}\right), T_{i_0}\left(x_{m(k)-1}\right)\right).\end{aligned} \tag{2.14}$$

Taking $k \to \infty$ in (2.14) and using (2.6), (2.7) and (2.13) we get

$$\lim_{k\to\infty} p\left(T_{i_0}\left(x_{n(k)}\right), T_{i_0}\left(x_{m(k)-1}\right)\right) = \epsilon. \tag{2.15}$$

Now using inequality (2.1), we have

$$\begin{aligned}\psi(\epsilon) &\le \psi p\left(T_{i_0}\left(x_{n(k)}\right), T_{i_0}\left(x_{m(k)}\right)\right)\\
&\le \varphi\left(M\left(x_{n(k)}, x_{m(k)}\right)\right),\end{aligned} \tag{2.16}$$

Where

$$M\left(x_{n(k)},x_{m(k)}\right)$$

$$=\max\left\{\begin{array}{l} p\left(g\left(x_{n(k)}\right),g\left(x_{m(k)}\right)\right), \\ p\left(g\left(x_{n(k)}\right),T_{\alpha_0}\left(x_{n(k)}\right)\right), \\ p\left(g\left(x_{m(k)}\right),T_{i_0}\left(x_{m(k)}\right)\right), \\ \dfrac{\left[\begin{array}{l} p\left(g\left(x_{n(k)}\right),T_{i_0}\left(x_{m(k)}\right)\right) \\ +p\left(g\left(x_{m(k)}\right),T_{i_0}\left(x_{n(k)}\right)\right) \end{array}\right]}{2} \end{array}\right\}$$

$$=\max\left\{\begin{array}{l} p\left(T_{i_0}\left(x_{n(k)-1}\right),T_{i_0}\left(x_{m(k)-1}\right)\right), \\ p\left(T_{i_0}\left(x_{n(k)-1}\right),T_{i_0}\left(x_{n(k)}\right)\right), \\ p\left(T_{i_0}\left(x_{m(k)-1}\right),T_{i_0}\left(x_{m(k)}\right)\right), \\ \dfrac{\left[\begin{array}{l} p\left(T_{i_0}\left(x_{n(k)-1}\right),T_{i_0}\left(x_{m(k)}\right)\right) \\ +p\left(T_{i_0}\left(x_{m(k)-1}\right),T_{i_0}\left(x_{n(k)}\right)\right) \end{array}\right]}{2} \end{array}\right\}.$$

Letting $k \to \infty$ in the above inequality and using (2.6),(2.10), (2.13) and (2.15), we obtain

$$\lim_{k\to\infty} M\left(x_{n(k)},x_{m(k)}\right)=\max\left\{\epsilon,0,0,\frac{\epsilon+\epsilon}{2}\right\}=\epsilon.$$

As $k \to \infty$, inequality (2.16) becomes,

$$\psi(\epsilon)\le\varphi(\epsilon)$$

which is a contradiction by virtue of a property of $\varphi$.

Then, we deduce that $\{T_{i_0}(x_n)\}$ is a cauchy sequence in $(X, p^s)$. Hence, we have

$$\lim_{m,n\to\infty} p^s\left(T_{i_0}(x_m), T_{i_0}(x_n)\right) = 0.$$

Now, from the definition of $p^s$ and from (2.7), we have

$$\lim_{m,n\to\infty} p\left(T_{i_0}(x_m), T_{i_0}(x_n)\right) = 0. \tag{2.17}$$

Since $X$ is a complete partial metric space, then, from lemma 1.5, the sequence $\{T_{i_0}(x_n)\}$ converges in the metric space $(X, p^s)$, so there exist $z$ in $X$ such that

$$\lim_{n\to\infty} p^s\left(T_{i_0}(x_n), z\right) = \lim_{n\to\infty} p^s\left(gx_{n+1}, z\right) = 0.$$

Again, from Lemma 1.5, we get

$$p(z,z) = \lim_{n\to\infty} p\left(T_{i_0}(x_n), z\right)$$
$$= \lim_{n\to\infty} p\left(gx_{n+1}, z\right) = \lim_{n,m\to\infty} p\left(T_{\alpha_0}(x_n), T_{i_0}(x_m)\right). \tag{2.18}$$

But, from (2.17) and (2.18) we have

$$p(z,z) = \lim_{n\to\infty} p\left(T_{i_0}(x_n), z\right) = \lim_{n\to\infty} p\left(gx_{n+1}, z\right) = 0. \tag{2.19}$$

Since $\{T_{i_0}(x_n)\} \subseteq gX$ and gX is closed, there exists x∈ X such that z = gx. Now, we claim that x is a coincidence point of Ti0 and g. We have

$$p\left(gx, T_{i_0}(x)\right)$$
$$\leq p\left(gx, gx_{n+1}\right) + p\left(gx_{n+1}, T_{i_0}(x)\right) - p\left(gx_{n+1}, gx_{n+1}\right)$$
$$= p\left(z, gx_{n+1}\right) + p\left(T_{i_0}(x_n), T_{i_0}(x)\right) - p\left(gx_{n+1}, gx_{n+1}\right).$$

On taking limit as $n \to \infty$ in the above inequality, we have

$$p\left(gx, T_{i_0}(x)\right) \leq \lim_{n\to\infty} p\left(T_{i_0}(x_n), T_{i_0}(x)\right), \tag{2.20}$$

By property of $\psi$ and using (2.20), we have

$$\psi\left(p\left(gx,T_{i_0}(x)\right)\right)\leq \lim_{n\to\infty}\psi\left(p\left(T_{i_0}(x_n),T_{i_0}(x)\right)\right)$$
$$\leq \lim_{n\to\infty}\varphi\left(M(x_n,x)\right)=\varphi\left(p\left(gx,T_{i_0}(x)\right)\right). \tag{2.21}$$

Indeed,

$$p\left(gx,T_{i_0}(x)\right)\leq M(x_n,x)$$
$$=\max\left\{\begin{array}{l} p(gx_n,gx),p\left(gx_n,T_{i_0}(x_n)\right),p\left(gx,T_{i_0}(x)\right),\\ \dfrac{p\left(gx_n,T_{i_0}(x)\right)+p\left(gx,T_{i_0}(x_n)\right)}{2}\end{array}\right\}$$
$$\leq\max\left\{\begin{array}{l} p(gx_n,gx),p\left(gx_n,T_{i_0}(x_n)\right),p\left(gx,T_{i_0}(x)\right),\\ \dfrac{\left[\begin{array}{l} p(gx_n,gx)+p\left(gx,T_{i_0}(x)\right)\\ -p(gx,gx)+p\left(gx,T_{i_0}(x_n)\right)\end{array}\right]}{2}\end{array}\right\}$$

we deduce, taking limit as $n\to\infty$, that

$$\lim_{n\to\infty}M(x_n,x)=p\left(gx,T_{i_0}(x)\right).$$

Now suppose that $x\in X$ is a coincidence point of $T_{i_0}$ and g. Then for any $i\in I$, from (2.1), we have

$$\psi p\left(gx,T_i(x)\right)=\psi\left(pT_{i_0}(x),T_i(x)\right)\leq\varphi\left(M(x,x)\right)$$

Where

$$M(x,x)=\max\left\{\begin{array}{l} p(g(x),g(x)),p\left(g(x),T_{i_0}(x)\right),\\ p\left(g(x),T_i(x)\right)\\ \dfrac{p\left(g(x),T_i(x)\right)+p\left(g(x),T_{i_0}(x)\right)}{2}\end{array}\right\}$$
$$=p\left(g(x),T_i(x)\right).$$

So,

$$\psi p\left(gx, T_i(x)\right) \leq \varphi p\left(g(x), T_i(x)\right),$$

therefore p(gx, $T_i(x)$) = 0, which implies that $T_i(x)$ = gx, that is, x is a coincidence point of $\{T_i : i \in I\}$ and g.

Assume that z, y are coincidence points of $\{T_i : i \in I\}$ and g in X such that y ≠ z. Then there exist $t_1$, $t_2$ in Xsuch that, for all i ∈ I, $T_i(t_1) = g(t_1) = z$ and $T_i(t_2) = g(t_2) = y$. Using (2.1), we have

$$\psi\left(p\left(gt_1, gt_2\right)\right) = \psi\left(p\left(T_{i_0}(t_1), T_i(t_2)\right)\right) \leq \varphi\left(M(t_1, t_2)\right)$$

Where

$$M(t_1, t_2) = \max\left\{\begin{array}{l} p(gt_1, gt_2), p\left(gt_1, T_{i_0}(t_1)\right), p\left(gt_2, T_i(t_2)\right) \\ \dfrac{p\left(gt_1, T_i(t_2)\right) + p\left(gt_2, T_{i_0}(t_1)\right)}{2} \end{array}\right\} = p\left(gt_1, gt_2\right).$$

So

$$\psi\left(p\left(gt_1, gt_2\right)\right) \leq \varphi\left(p\left(gt_1, gt_2\right)\right) < \psi\left(p\left(gt_1, gt_2\right)\right),$$

which is a contradiction which proves the uniqueness of point of coincidence.

**Example 2.2.** Let X = [0, 1] and p(x, y) = max{x, y}, then it is clear that (X, p) is a complete partial metric space. Let I = {1, 2, 3, ...}, i0 = 1 and for every i ∈ I, Ti, g : X → X, and ψ, φ: [0,∞) → [0,∞) be given by $T_i(x) = \frac{x^3}{(2i+1)(x+3)}$, $gx = \frac{x^2}{2x+6}$, $\psi(t) = 2t$, $\varphi(t) = \frac{4t}{3}$. Clearly (ψ, φ) ∈ Ψ. We show that condition (2.1) is satisfied.

If x, y ∈ X, then we have

$$\psi p\left(T_{i_0}(x),T_i(y)\right)=\psi\left(p\left(T_1(x),T_i(y)\right)\right)$$

$$=\psi\left(\max\left\{\frac{x^3}{3x+9},\frac{y^3}{(2i+1)(3y+9)}\right\}\right)$$

$$\leq 2\max\max\left\{\frac{x^3}{3x+9},\frac{y^3}{(2i+1)(3y+9)}\right\}$$

$$\leq\frac{2}{3}\max\left\{\frac{x^3}{x+3},\frac{y^3}{(y+3)}\right\}=\frac{4}{3}p(gx,gx)$$

$$\leq\frac{4}{3}\max\left\{\begin{array}{l}p(gx,gx),p(gx,T_1(x)),p(gy,T_i(y)),\\ \dfrac{p(gx,T_i(y)),p(gy,T_1(x))}{2}\end{array}\right\}$$

$$=\varphi(M(x,y)).$$

*Note that, $\{T_i\}$ and g satisfy all the conditions given in Theorem 2.1. Moreover, 0 is a unique common fixed point of $\{T_i\}$ and g.*

As immediate consequences of Theorem 2.1 are the following fixed point results.

**Corollary 2.3.** Let (X, p) be a complete partial metric space. Suppose $T_1$, $T_2$, $g : X \to X$ be three self mappings such that $T_1(x)\subseteq g(X)$, gX is closed in (X, p) and for all x, y ∈ X

$$\psi\left(p(T_1x,T_2y)\right)\leq\varphi\left(M(x,y)\right),$$

*Where*

$$M(x,y)=\max\left\{\begin{array}{l}p(gx,gy),p(gx,T_1x),p(gy,T_2y),\\ \dfrac{p(gx,T_2y),p(gy,T_1x)}{2}\end{array}\right\},$$

and $(\psi, \varphi)\in\Psi$. Then, there exists a unique x ∈ X such that $gx = T_1x = T_2x$. Moreover, we have p(gx, gx) = 0.

**Proof**. Considering $\{T_I : I \in I\} = \{T_1, T_2\}$ in Theorem 2.1, we have the required proof.

**Remark 2.4.** *Corollary 2.3 extends and generalizes many existing fixed point theorems.*

If we take $\psi(t) = t$ and $T_i = T$ in Theorem 2.1, we have the following corollary.

**Corollary 2.5.** Let (X, p) be a complete partial metric space. Suppose T: X → X be a self mapping such that $T(X) \subseteq g(X)$, gX is closed in (X, p) and for all x, y ∈ X

$$p(Tx, Ty) \le \varphi(M(x, y)),$$

*where φ: [0,∞) → [0,∞) is continuous, nondecreasing function such that ϕ(t) < t for each t > 0 and*

$$M(x,y) = \max\left\{ \begin{matrix} p(gx,gy), p(gx,Tx), p(gy,Ty), \\ \dfrac{p(gx,Ty), p(gy,Tx)}{2} \end{matrix} \right\}.$$

Then, there exists a unique x ∈ X such that gx = Tx. Moreover, we have p(gx, gx) = 0.

If we take g = I and φ(t) = λ t for λ ∈ [0, 1) in Corollary 2.5, we have the following corollary.

**Corollary 2.6.** Let (X, p) be a complete partial metric space. Suppose T : X → X be a self mapping and g: X→X. Also $T(X) \subseteq g(X)$, gX is closed in (X, p) and for all x, y ∈ X

$$p(Tx,Ty) \le \lambda \max\left\{ \begin{matrix} p(x,y), p(x,Tx), p(y,Ty), \\ \dfrac{p(x,Ty), p(y,Tx)}{2} \end{matrix} \right\}.$$

Then, there exists a unique x ∈ X such that x = Tx. Moreover, we have p(x, x) = 0.

Now, we will prove the following result.

**Theorem 2.7.** Let (X, p) be a complete partial metric space. Suppose $\{T_i : X \to X : i \in I\}$ be a family of mappings and $g : X \longrightarrow X$ be a self map. Also there exists $i_0 \in I$ such that $T_{i_0}(X) \subseteq g(X)$ and

$$\psi\left(p\left(T_{i_0}x, T_i y\right)\right) \le \varphi\left(M(x,y)\right), \tag{2.22}$$

*Where*

$$M(x,y) = \max\left\{ \begin{array}{c} p(gx,gy), p\left(gx, T_{i_0}x\right), p(gy, T_i y), \\ \dfrac{p(gx, T_i y), p\left(gy, T_{i_0}x\right)}{2} \end{array} \right\}, \tag{2.23}$$

and $(\psi, \varphi) \in \Psi$. Also suppose $T_{i_0}$, g are partial compatible and continuous in (X, p). Then, there exists a unique $x \in X$ such that $gx = T_i x$ for all $i \in I$. Moreover, we have $p(gx, gx) = 0$.

**Proof**. Using the same arguments in the proof of Theorem 2.1, we deduce that $\{T_{i_0}(x_n)\}$ is a Cauchy sequence in the complete metric space $(X, p^s)$, and therefore, there exists $z \in X$ such that

$$\begin{aligned} p(z,z) &= \lim_{n\to\infty} p\left(T_{i_0}(x_n), z\right) = \lim_{n\to\infty} p\left(gx_{n+1}, z\right) \\ &= \lim_{n,m\to\infty} p\left(T_{i_0}(x_n), T_{i_0}(x_m)\right) = 0. \end{aligned} \tag{2.24}$$

Now we show that *z* is a coincidence point of $T_{i_0}$ and *g*.

Since $T_{i_0}$ and *g* are continuous, from (2.24) and using Lemma 1.7, we get

$$p\left(T_{i_0}z, T_{i_0}z\right) = \lim_{n\to\infty} p\left(T_{i_0}\left(T_{i_0}(x_n)\right), T_{i_0}z\right),$$

$$p(gz,gz)=\lim_{n\to\infty} p\left(g\left(T_{i_0}(x_{n+1})\right),gz\right),$$

(2.25)Since $T_{i_0}$ and $g$ are partial compatible mappings, this implies that

$$p(gz,gz)=0$$
$$and\ \lim_{n\to\infty} p\left(g\left(T_{i_0}(x_{n+1})\right),g\left(T_{i_0}(x_{n+1})\right)\right)=0. \tag{2.26}$$

The condition $(p_4)$, we obtain

$$\begin{aligned}
p\left(T_{i_0}z,gz\right) &\le p\left(T_{i_0}z,T_{i_0}\left(T_{i_0}(x_n)\right)\right)\\
&+p\left(T_{i_0}\left(T_{i_0}(x_n)\right),g\left(T_{i_0}(x_{n+1})\right)\right)\\
&+p\left(g\left(T_{i_0}(x_{n+1})\right),gz\right)\\
&-p\left(T_{i_0}\left(T_{i_0}(x_n)\right),T_{i_0}\left(T_{i_0}(x_n)\right)\right)\\
&-p\left(g\left(T_{i_0}(x_{n+1})\right),g\left(T_{i_0}(x_{n+1})\right)\right)\\
&= p\left(T_{i_0}z,T_{i_0}\left(T_{i_0}(x_n)\right)\right)\\
&+p\left(T_{i_0}(gx_{n+1}),g\left(T_{i_0}(x_{n+1})\right)\right)\\
&+p\left(g\left(T_{i_0}(x_{n+1})\right),gz\right)\\
&-p\left(T_{i_0}\left(T_{i_0}(x_n)\right),T_{i_0}\left(T_{i_0}(x_n)\right)\right)\\
&-p\left(g\left(T_{i_0}(x_{n+1})\right),g\left(T_{i_0}(x_{n+1})\right)\right).
\end{aligned}$$

Letting $n\to\infty$ in the above inequality and using (2.24), (2.25) and (2.26), we have

$$\begin{aligned}
p\left(T_{i_0}z,gz\right) &\le p\left(T_{i_0}z,T_{i_0}z\right)+p(gz,gz)\\
&= p\left(T_{i_0}z,T_{i_0}z\right).
\end{aligned} \tag{2.27}$$

Now, we will prove that $p(T_{i_0}z, T_{i_0}z) = 0$. Suppose that this is not the case. Then, from (2.22) with $i = i_0$, $y = x = z$, we get

$$\psi\left(p\left(T_{i_0}z, T_{i_0}z\right)\right) \le \varphi(M(z,z)),$$

Where

$$M(z,z) = \max\left\{\begin{array}{l} p(gz,gz), p\left(gz, T_{i_0}z\right), p\left(gz, T_{i_0}z\right), \\ \dfrac{p\left(gz, T_{i_0}z\right), p\left(gz, T_{i_0}z\right)}{2} \end{array}\right\}$$
$$= p\left(gz, T_{\alpha_0}z\right).$$

Therefore, from (2.27) and the above inequality, we have

$$p\left(gz, T_{i_0}z\right) < p\left(gz, T_{i_0}z\right),$$

a contradiction. Hence $p(gz, T_{i_0}z) = 0$ which implies that $T_{i_0}z = gz$, that is, $z$ is a coincidence point of $T_{i_0}$ and $g$. If $w$ is a coincidence point of $T_{i_0}$ and $g$ other then $z$, then putting $i = i_0$, $x = z$, $y = w$ in (2.22) we have

$$\psi\left(p\left(T_{i_0}z, T_{i_0}w\right)\right) \le \varphi(M(z,w)),$$

Where

$$M(z,w) = \max\left\{\begin{array}{l} p(gz,gw), p\left(gz, T_{i_0}z\right), p\left(gw, T_{i_0}w\right), \\ \dfrac{p\left(gz, T_{i_0}w\right), p\left(gw, T_{i_0}z\right)}{2} \end{array}\right\}$$
$$= \max\left\{\begin{array}{l} p\left(T_{i_0}z, T_{i_0}w\right), p\left(T_{i_0}z, T_{i_0}z\right), p\left(T_{\alpha_0}w, T_{\alpha_0}w\right), \\ \dfrac{p\left(T_{i_0}z, T_{i_0}w\right) + p\left(T_{i_0}w, T_{i_0}z\right)}{2} \end{array}\right\}$$
$$= p\left(T_{i_0}z, T_{i_0}w\right).$$

Therefore,

$$\psi\left(p\left(T_{i_0}z,T_{i_0}w\right)\right)\le\varphi\left(p\left(T_{i_0}z,T_{i_0}w\right)\right),$$

which is possible only when z = w. Hence z is the unique coincidence point of $T_{i_0}$ and g. Similar Theorem 2.1 can be shown that z is a coincidence point of $\{T_i\}$ and g for any i ∈ I.

We give in the following a sufficient condition for the uniqueness of the common fixed point of the mappings *$\{T_i\}$* and *g*.

**Theorem 2.8.** *Adding to the hypotheses of Theorem 2.1 (resp. Theorem 2.7) the condition:*

*$T_{i_0}$ and g commute at their coincidence points, we obtain the uniqueness of common fixed point of $\{T_i\}$ and g.*

**Proof**. Suppose that $T_{i_0}$ and *g* commute at *x*. Set $y = T_{i_0}x = gx$. Then

$$T_{i_0}y=T_{i_0}(gx)=g\left(T_{i_0}x\right)=gy, \tag{2.28\_}$$

from (2.1) we get

$$\begin{aligned}\psi\left(p\left(y,T_{i_0}y\right)\right)&=\psi\left(p\left(T_{i_0}x,T_{i_0}y\right)\right)\\&\le\psi\left(M\left(x,y\right)\right),\end{aligned} \tag{2.29}$$

Where

$$M(x,y) = \max\left\{ \begin{array}{l} p(gx,gy), p(gx,T_{i_0}x), p(gy,T_{i_0}y), \\ \dfrac{p(gx,T_{i_0}y), p(gy,T_{i_0}x)}{2} \end{array} \right\}$$

$$= \max\left\{ \begin{array}{l} p(T_{i_0}x,T_{i_0}y), p(T_{i_0}x,T_{i_0}x), p(T_{i_0}y,T_{i_0}y), \\ \dfrac{p(T_{i_0}x,T_{i_0}y) + p(T_{i_0}y,T_{i_0}x)}{2} \end{array} \right\}$$

$$= p(T_{i_0}x,T_{i_0}y) = p(y,T_{i_0}y).$$

Suppose that $p(y, T_{i_0} y) > 0$, from (2.29), we get

$$\psi(p(y,T_{i_0}y)) \leq \varphi(M(x,y))$$
$$= \varphi(p(y,T_{i_0}y)) < \psi(p(y,T_{i_0}y)),$$

which is a contradiction. Hence $p(y, T_{\alpha_0} y) = 0$. Therefore,

$$T_{i_0}y = gy = y. \tag{2.30}$$

Thus we proved that $T_{i_0}$ and $g$ have a common fixed point.

Uniqueness: Let $v$ and $w$ be two common fixed points of $T_{i_0}$ and $g$. (i.e) $v = T_{i_0} v = gv$ and $w = T_{i_0} w = gw$. Using inequality (2.1), we have

$$\psi(T_{i_0}w, T_{i_0}v) \leq \varphi(M(w,v)),$$

Where

$$M(w,v) = \max\left\{ \begin{array}{l} p(gw,gv), p(gw,T_{i_0}w), p(gv,T_{i_0}v), \\ \dfrac{p(gw,T_{i_0}v) + p(gv,T_{i_0}w)}{2} \end{array} \right\}$$
$$= p(w,v).$$

Therefore,

$$\psi(p(w,v)) = \psi\left(p\left(T_{i_0}w, T_{i_0}v\right)\right) \leq \varphi(p(w,v))$$

which is possible only when $w = v$. Hence $T_{i_0}$ and $g$ have an unique common fixed point. Now suppose that $x \in X$ is a common fixed point of $T_{i_0}$ and $g$. Then $x = gx = T_{i_0}x$.

For any i ∈ I, from inequality (2.1), we have

$$\psi(p(x,T_i x)) = \psi\left(p\left(T_{i_0}x, T_i x\right)\right) \leq \varphi(M(x,x)) \tag{2.31}$$

Where

$$M(x,x) = \max\left\{\begin{matrix} p(gx,gx), p(gx,T_i x), p\left(gx, T_{i_0}x\right), \\ \dfrac{p(gx,T_i x) + p(x,x)}{2} \end{matrix}\right\}$$

$$= \max\left\{\begin{matrix} p(x,x), p(x,T_i x), p(x,x), \\ \dfrac{p(x,T_i x) + p(x,x)}{2} \end{matrix}\right\}$$

$$= p(x,T_i x)$$

Suppose that $p(x, T_i x) > 0$, from (2.31), we get

$$\psi(p(x,T_i x)) \leq \varphi(M(x,x))$$
$$= \varphi(p(x,T_i x)) < \psi(p(x,T_i x)),$$

which is a contradiction. Hence $p(x, T_i x) = 0$. So, we have

$$T_i x = x.$$

that is, $x$ is a common fixed point of $g$ and $\{T_i : i \in I\}$.

**Example 2.9.** Let $X = [0, 1]$ and $p(x, y) = \max\{x, y\}$, then it is clear that $(X, p)$ is a complete partial metric space. Let $I = \{1, 2, 3, ...\}$, $i_0 = 1$ and for every $i \in I$, $T_i$, $g : X \to X$, and $\psi$, $\varphi : [0,\infty) \to [0,\infty)$ be given by $T_i(x)=\frac{10}{i+2}x^3$, $gx=4x^3$, $\psi(t)=4t$, $\varphi(t)=\frac{10}{3}t$, Clearly $(\psi, \varphi) \in \Psi$. We show that condition (2.1) is satisfied.

If x, y ∈ X, then we have

$$\psi\left(p\left(T_{i_0}(x),T_i(y)\right)\right)=\psi\left(p\left(T_1(x),T_i(y)\right)\right)$$
$$=\psi\left(\max\left\{\frac{10}{3}x^3,\frac{10}{i+2}y^3\right\}\right)$$
$$=4\max\left\{\frac{10}{3}x^3,\frac{10}{i+2}y^3\right\}$$
$$\leq\frac{4}{3}\max\left\{10x^3,10y^3\right\}=\frac{10}{3}p(gx,gy)$$
$$\leq\frac{10}{3}\max\left\{\begin{array}{l}p(gx,gy),p(gx,T_1(x)),p(gy,T_i(y)),\\ \frac{p(gx,T_i(y)),p(gy,T_1(x))}{2}\end{array}\right\}$$
$$=\varphi(M(x,y)).$$

*Note that, $\{T_i\}$ and g satisfy all the conditions given in Theorem 2.7 and 2.8. Moreover, 0 is a unique commonfixed point of $\{T_i\}$ and g.*

# REFERENCES


1. T. Abdeljawad, E. Karapinar, K. Tas, Existence and uniqueness of a common fixed point on partial metric spaces, Applied Mathematics Letters, 24(11)(2011),1900-1904.

2. Altun, A. Erduran, Fixed point theorems for monotone mappings on partial metric spaces, Fixed Point Theory Appl 2011, 10.

3. Altun, F. Sola, H. Simsek, Generalized contractions on Partial metric spaces, Topology and its Applications. 157 (2010) 2778-2785.

4. Altun, H. Simsek, Some fixed point theorems on dualistic partial metric spaces, J Adv Math Stud. 1, (2008) 1-8.

5. Altun, A. Erduran, Fixed point theorems for monotone mappings on partial metric spaces, Fixed Point Theory Appl. 2011.

6. M. Bukatin, R. Kopperman, S. Matthews, H. Pajoohesh, Partial metric spaces, Am. Math. Mon. 116,(2009) 708-718.

7. E. Karapinar, M. E. Inci, Fixed point theorems for operators on partial metric spaces, Appl Math Lett 24 (11)(2011) 1894-1899.

8. S. G. Matthews, Partial metric topology, in: Proc. 8th Summer Conference on General Topology and Applications, in: Ann. New York Acad. Sci., vol. 728, 1994, pp. 183-197.

9. S. Oltra, O. Valero, Banachs fixed point theorem for partial metric spaces, Rend. Istit. Mat. Univ. Trieste 36 (2004) 17-26.

10. S. J. ONeill, Two topologies are better than one, Tech. report, University of Warwick, Coven-try, UK, http://www.dcs.warwick.ac.uk/reports/283.html, 1995.

11. S. J. ONeill, Partial metrics, valuations and domain theory, in: Proc. 11th Summer Conference on General Topology and Applications, in: Ann. New York Acad. Sci., vol. 806, 1996, pp. 304-315.

12. S. Romaguera, A Kirk type characterization of completeness for partial metric spaces, Fixed Point Theory Appl. 2010 (Ar-ticle ID 493298),6 (2010).

13. S. Romaguera, O. Valero, A quantitative computational model for complete partial metric spaces via formal balls, Math. Structures Comput. Sci. 19 (3) (2009), 541-563.

14. B. Samet, M. Rajovic, R. Lazovi, R. Stoiljkovic, Common Fixed Point Results For Nonlinear Contractions in Ordered Partial Metric Spaces, Fixed Point Theory Appl. 2011.

15. O. Valero, On Banach fixed point theorems for partial metric spaces, Appl. Gen. Topol, 6 (2) (2005) 229-240.

6

# DESCENT VIA (5,5)-ISOGENY ON JACOBIANS OF GENUS 2 CURVES

**E.V. Flynn**

Mathematical Institute, University of Oxford, Andrew Wiles Building, Woodstock Road, Oxford OX2 6GG, United Kingdom

## ABSTRACT

We describe a family of curves CC of genus 2 with a maximal isotropic $(Z/5)^2$(Z/5)2 in J[5]J[5], where $J$ is the Jacobian variety of CC, and develop the theory required to perform descent via (5,5)(5,5)-isogeny. We apply this to several examples, where it can be shown that non-reducible Jacobians have nontrivial 5-part of the Tate–Shafarevich group.

## KEYWORDS

Higher genus curves; Jacobians; Tate–Shafarevich group

## 1. INTRODUCTION

In this article we construct a family of curves CC of genus 2 over a field $k$ of characteristic not 2 or 3, with Jacobian $J$ , where J[5](k¯)

contains a group Σ of order 25 which is defined over *k*. As we show in Section 2, when C has the form

$$y^2 = F(x) = \lambda G(x)^2 + H(x)^5, \tag{1}$$

where G(x) is quadratic and H(x) is linear in $x$ , then the divisor $D = [(x_0, \sqrt{\lambda}G(x_0)) - \infty]$ where $x_0$x0 is the root of H(x) and ∞ is the point at infinity on the curve, represents a point in $J[5](\bar{k})$ as can be seen from the fact that $5D$ is linearly equivalent to the divisor of the function $y - \sqrt{\lambda}G(x)$ Furthermore the group generated by *D* is defined over *k*.

In Section 2, we shall describe curves which can be written in the form (1) in two ways, such that $\Sigma \subset J[5]$ is maximally isotropic with respect to the Weil pairing, so that $\tilde{J}=J/\Sigma$ is also principally polarized and in fact has a $\Sigma^\vee$-level structure. In Section 3, we shall give an explicit family of such curves and in Section 4 we identify a curve $\tilde{C}$ such that its Jacobian $\tilde{J}_r$ is $J_r/\Sigma$. This gives rise to a (5,5)-isogeny $\phi: J_r \mapsto \tilde{J}_r$, defined over $k$ , and a dual isogeny $\tilde{\phi}: \tilde{J}_r \mapsto J_r$ such that $\tilde{\phi} \circ \phi = [5]$. The explicit derivation of the model of $\tilde{C}_r$ and the isogeny $\phi$ will require some significant computation on the Kummer surface.

In Section 5, we describe explicitly how to perform a descent via (5,5)(5,5)-isogeny, which allows us to compute a bound on the rank on J(k)J(k), and to obtain information about Ш(J/k)[5]Ш(J/k)[5], using ideas from [3] and [6]. This allows us to give an example of an absolutely simple abelian surface with nontrivial elements of Ш(J/Q)[5]Ш(J/Q)[5]; as far as we are aware, this is the first such known example.

The broad strategy will be similar to that developed for (3,3)(3,3)-isogenies in [2], using maps on the Kummer surface, but with considerably greater computational complexity, due to the larger dimensional space of quintics required here and the fact that, for our family, the members of the kernel are not individually defined over Q. We shall also, in our second example, use a finesse in which a nontrivial element of Ш(J/Q)[5] is exhibited using the easier direction of the (5,5)(5,5)-isogeny which requires only quadratic number fields for the descent map.

## 2. FIVE-TORSION ON GENUS TWO JACOBIANS

Let $k$ be a field of characteristic different from 2, 3 and let CC be a smooth projective curve of genus 2 over $k$ given by an affine model

$$\mathcal{C}: y^2 = F(x),$$

where $F(x) \in k[x]F(x) \in k[x]$ is of degree 5 or 6 (this only mildly restricts the admissible CC if $k$ is a finite field of 5 elements). Let $J$ be the Jacobian of CC. The group $J(k)J(k)$ is isomorphic to the group of divisor classes of $k$ -rational degree 0 divisors on CC. Since CC is of genus 2, every degree 0 class contains a representative of the form

$$D - \kappa,$$

where $D$ is an effective divisor of degree 2 and $\kappa$ is an effective canonical divisor. Furthermore, for any non-principal class, the divisor $D$ is unique. In the following, our curves will all have $F(x)F(x)$ of degree 5 and we shall take $\kappa=2\infty\kappa=2\infty$.

**Lemma 1.** *Let* CC*be a curve of genus* 2 *of the form*

$$\mathcal{C}: y^2 = F(x) = \lambda_1 G_1(x)^2 + H_1(x)^5 = \lambda_2 G_2(x)^2 + H_2(x)^5,$$

*where* $H_1, H_2, G_1, G_2, F \in k[x]$, $\lambda_1, \lambda_2 \in k^\times$, $G_1$, $G_2$ *are of degree* 2, $H_1$, $H_2$*are of degree* 1 *and* $\gcd(H_1,H_2)=1$gcd(. *Then* $\mathrm{Pic}(\mathcal{C}/\bar{k})[5]$ *has a subgroup* $\Sigma$ *of size* 25 *which is defined over k.*

**Proof.** For each $i=1,2$, let $x_i$ be the root of $H_i$Hi, and let $T_i = [(x_i, \sqrt{\lambda_i}G_i(x_i)) - \infty]$. By considering the divisor of the function

$y-\sqrt{\lambda_i}G_i(x_i)$ we see that $5T_i=0$, the identity, and $T_iTi$ has order 5. Furthermore, $\langle T_i$, the group generated by $T_i$, is:

$$0T_i = \mathcal{O},$$
$$1T_i = [(x_i, \sqrt{\lambda_i}G_i(x_i)) - \infty],$$
$$2T_i = [2(x_i, \sqrt{\lambda_i}G_i(x_i)) - 2\infty],$$
$$3T_i = [2(x_i, -\sqrt{\lambda_i}G_i(x_i)) - 2\infty],$$
$$4T_i = [(x_i, -\sqrt{\lambda_i}G_i(x_i)) - \infty],$$

so that $\langle T_i\rangle$ is defined over *k* . Furthermore $T_2 \notin \langle T_1\rangle$ , so that $\langle T_1,T_2\rangle$ is of order 25 and is defined over *k*. □

The torsion subgroup scheme J[5] comes equipped with a non-degenerate, bilinear, alternating *Weil pairing*

$$e_5\colon J[5] \times J[5] \to \mu_5,$$

where $\mu_5\mu5$ is the group scheme representing the fifth roots of unity.

We say that a subgroup $\Sigma \subset J[5]$ is *isotropic* if $e_5e5$ restricts to the trivial pairing on Σ. If Σ is of degree 25 then Σ is *maximal isotropic* , meaning Σ is not properly contained in an isotropic subgroup. From the properties of the Weil pairing, there is an induced isomorphism $J[5]/\Sigma \to \Sigma^\vee = \mathrm{Hom}(\Sigma,\mu_5)$. In fact, we have a direct sum decomposition $J[5]=\Sigma\times\Sigma^\vee$. In particular, if $\Sigma=Z/5\times Z/5$ then we have $\Sigma^\vee=\mu_5\times\mu_5$. The isotropy guarantees that $J/\Sigma J/\Sigma$ is principally polarized.

Ed Schaefer has communicated the following lemma and proof, which gives the required property of the Weil pairing in our case.

**Lemma 2.** *Let p be an odd prime, k be a field of characteristic* 0, $C/kC/k$*be a curve of genus greater than* 1, $C(k)\neq\emptyset$, *J be its Jacobian,* $\alpha,\beta\in J[p]$ *and* $\mu_p\not\subseteq K(\alpha,\beta)$ *the minimal field of definition of α and β over k). Then* $e_p(\alpha,\beta)=1$

**Proof.** Since $C(k)\neq\emptyset$ there are degree 0 divisors $D_\alpha$ and $D_\beta$, defined over $K(\alpha)$ and $K(\beta)$ respectively, with disjoint supports, such that $\alpha=[D_\alpha]$ and $\beta=[D_\beta]$. There are also functions $f_\alpha$ and $f_\beta$, defined over $K(\alpha)$ and $K(\beta)K(\beta)$ respectively, such that $divis(f_\alpha)= pD_\alpha divis(f\alpha)=pD\alpha$ and $divis(f_\beta)=pD_\beta divis(f\beta)=pD\beta$. Using the Weil reciprocity definition of the Weil pairing on a Jacobian, we have $e_p(\alpha,\beta)=f_\beta(D_\alpha)/f_\alpha(D_\beta)$ and $e_p(\alpha,\beta)\in K(\alpha,\beta)\cap\mu_p=\{1\}$. □

The above motivates that we should expect to be able to find an isogenous Jacobian. In Sections 3, 4 we shall find a family of curves $C_r$ and $\widetilde{C}_r$, and the computation in Section 4 will show independently that there is an isogeny between the Jacobians $J_r Jr$ and $\widetilde{J}_r$ with kernel $\Sigma$.

## 3. A FAMILY OF GENUS 2 CURVES WITH A MAXIMAL ISOTROPIC $(Z/5)^2$ IN J[5]

Let $k$ be a field of characteristic different from 2, 3 and let $r\in k$. In this section we derive a genus 2 curve $C_r$ over $k$ with two non-trivial divisor classes $T_1,T_2\in Pic(C_r/k)[5]$ with $T_1\notin\langle T_2\rangle$ $T1\notin\langle T2\rangle$ and $e_5(T_1,T_2)=1e5(T1,T2)=1$.

In order for CC to have the form given in Lemma 1, we shall apply a linear transformation to map the roots of $H_1$, $H_2$ to $r$ , 0, respectively, and assume $H_1(x)=x-r$, $H_2(x)=x$. Then $\lambda_1G_1(x)^2+H_1(x)^5= \lambda_2G_2(x)^2+ H_2(x)^5$. Let $F_i(x)=\sqrt{\lambda_i}G_i(x)$ so that $F_1(x)^2+H_1(x)^5=F_2(x)^2+H_2(x)^5$ Then

$$\big(F_1(x)+F_2(x)\big)\big(F_1(x)-F_2(x)\big)=x^5-(x-r)^5=w_1w_2,$$

where

$$w_1=\frac{\sqrt{r}}{2}((5+\sqrt{5})x^2-r(5+\sqrt{5})x+2r^2),$$

$$w_2=\frac{\sqrt{r}}{2}((5-\sqrt{5})x^2-r(5-\sqrt{5})x+2r^2).$$

A natural approach here would be to solve $F_1$, $F_2$ for $F_1+F_2=w_1$, $F_1-F_2=w_2$, that is, $F_1=(w_1+w_2)/2$ and $F_2=(w_1-w_2)/2$. This gives:

$$F_1(x) = \frac{\sqrt{r}}{2}(5x^2 - 5rx + 2r^2), \quad F_2(x) = \frac{\sqrt{5r}}{2}x(x-r).$$

These are each of the required type, and they satisfy the required $F_1(x)^2+(x-r)^5=F_2(x)^2+x^5$, but $y^2=F_1(x)^2+(x-r)^5=F_2(x)^2+x^5$ has a repeated root at 0, and so is not of genus 2.

Instead we vary the above idea, by taking any unit $u$ in the ring of integers of $\mathbb{Q}(\sqrt{5})$, such as $u = (1+\sqrt{5})/2$, and now adjust by taking: $v_1=uw_1$ and $v_2=(1/u)w_2$v2=(1/u)w2. We now take $F_1 = (v_1+v_2)/2$ and $F_2 = (v_1 - v_2)/2$, so that $F_1^2 - F_2^2 = v_1v_2 = w_1w_2$. This gives

$$F_1(x) = \frac{\sqrt{5r}}{2}(3x^2 - 3rx + r^2), \quad F_2(x) = \frac{\sqrt{r}}{2}(5x^2 - 5rx + r^2).$$

So, now we have $y^2 = F_1(x)^2 + (x-r)^5 = F_2(x)^2 + x^5$ satisfying our requirements. In order to clear denominators, we replace *x*, *y* with x/4, y/2, giving

$$\begin{aligned} C_r : y^2 &= 5r(3x^2 - 12rx + 16r^2)^2 + (x-4r)^5 \\ &= r(5x^2 - 20rx + 16r^2)^2 + x^5. \end{aligned}$$

This can be summarized as follows.

**Lemma 3.** *Let $G_1(x) = 3x^2 - 12rx + 16r^2$, $G_2(x) = 5x^2 - 20rx + 16r^2$, $\lambda_1 = 5r$, $\lambda_2 = r$,*

*$H_1(x) = x - 4r$ and $H_2(x) = x$, for $r \in k^\times$. Then*

$$C_r\colon y^2 = F(x) = \lambda_1 G_1(x)^2 + H_1(x)^5 = \lambda_2 G_2(x)^2 + H_2(x)^5,$$

*has* $T_1 = [(4r, 16r^2\sqrt{5r}) - \infty]$, $T_2 = [(0, 16r^2\sqrt{r}) - \infty]$ *in* $J[5]$, *with* $\langle T_1, T_2\rangle$ *defined over k.*

We note also that the curve $C_1$C1 can be shown to have absolutely simple Jacobian, using the technique in [7] at the prime 19. Since this is a family of quadratic twists, they are geometrically the same curve, and so each $C_r$Cr has absolutely simple Jacobian.

## 4. THE ISOGENY

We consider the curve $C_r$ as defined in Lemma 3 and its Jacobian $J_r$Jr. In this section we determine a curve C˜r whose Jacobian $\widetilde{J}_r$r is isogenous to $J_r$Jr via an isogeny $\phi: J_r \to J_r/\Sigma$ where $\Sigma = \langle T_1, T_2\rangle$ . We do so by determining the corresponding map between their Kummer surfaces.

For CC a general curve of genus 2 given by a model

$$C: y^2 = f_6x^6 + f_5x^5 + \cdots + f_0,$$

with Jacobian *J*, we follow [4, p. 17] and choose a particular set of coordinates for the Kummer surface, given by $\xi = \xi(D) = \xi_0, \ldots, \xi_3$ as functions on *J* in terms of a divisor class $D = [(x_1, y_1) + (x_2, y_2) - \kappa]$ on CC as follows.

$$\xi_0 = 1, \quad \xi_1 = x_1 + x_2, \quad \xi_2 = x_1x_2, \quad \xi_3 = \frac{\Phi(\xi_0, \xi_1, \xi_2) - 2y_1y_2}{\xi_1^2 - 4\xi_0\xi_2}, \text{ where}$$
$$\Phi(\xi_0, \xi_1, \xi_2) = 2f_0\xi_0^3 + f_1\xi_0^2\xi_1 + 2f_2\xi_0^2\xi_2 + f_3\xi_0\xi_1\xi_2 + 2f_4\xi_0\xi_2^2 + f_5\xi_2^2\xi_1 + 2f_6\xi_2^3. \quad (2)$$

The quartic equation for the model of KK arising from these coordinates has the shape

$$\mathcal{K}: (\xi_1^2 - 4\xi_0\xi_2)\xi_3^2 + \Phi(\xi_0, \xi_1, \xi_2)\xi_3 + \Psi(\xi_0, \xi_1, \xi_2) = 0,$$

where $\Psi(\xi_0,\xi_1,\xi_2)$ is a quartic form we do not need explicitly here. The important observation is that one can read off the coefficients $f_0,\ldots,f_6$ directly from $\Phi$ and thus recover CC from it.

Let $C=C_r$. In order to produce $\Sigma$-invariant forms on $K_r$, we use biquadratic forms from [4, p. 23], arising from the addition structure on $J_r$Jr. For i,j=0,...,3i,j=0,...,3 we have forms

$$B_{i,j} \in k[\xi_0,\ldots,\xi_3,\xi'_0,\ldots,\xi'_3],$$

biquadratic in $(\xi_0,\ldots,\xi_3)$ and $(\xi'_0,\ldots,\xi'_3)$ such that for points $D_1$, $D_2$ on $J_r$Jr we have, as projective matrices,

equation(3)

$$\left(\xi_i(D_1+D_2)\,\xi_j(D_1-D_2)+\xi_i(D_1-D_2)\,\xi_j(D_1+D_2)\right)=\left(B_{ij}(\xi(D_1),\xi(D_2))\right).$$

Let $T_1,T_2\in J_r[5]$ be as in Lemma 3, that generate $\Sigma$, write $\xi(T_1),\xi(T_2)$ for the coordinate vectors of their images on the quartic model of $K_r$Kr and define

$$\begin{aligned}
R_{ij}(\xi_0,\ldots,\xi_3) &= B_{ij}(\xi_0,\ldots,\xi_3,\xi(T_1)),\\
S_{ij}(\xi_0,\ldots,\xi_3) &= B_{ij}(\xi_0,\ldots,\xi_3,\xi(T_2)),\\
R'_{ij}(\xi_0,\ldots,\xi_3) &= B_{ij}(\xi_0,\ldots,\xi_3,\xi(2T_1)),\\
S'_{ij}(\xi_0,\ldots,\xi_3) &= B_{ij}(\xi_0,\ldots,\xi_3,\xi(2T_2)).
\end{aligned} \tag{4}$$

We see that the quintic forms

$$
\begin{aligned}
R_{ijkl} = {} & \xi_i R_{jk} R'_{lm} + \xi_i R_{jl} R'_{km} + \xi_i R_{jm} R'_{kl} \\
& + \xi_i R_{kl} R'_{jm} + \xi_i R_{km} R'_{jl} + \xi_i R_{lm} R'_{jk} \\
& + \xi_j R_{ik} R'_{lm} + \xi_j R_{il} R'_{km} + \xi_j R_{im} R'_{kl} \\
& + \xi_j R_{kl} R'_{im} + \xi_j R_{km} R'_{il} + \xi_j R_{lm} R'_{ik} \\
& + \xi_k R_{ji} R'_{lm} + \xi_k R_{jl} R'_{im} + \xi_k R_{jm} R'_{il} \\
& + \xi_k R_{il} R'_{jm} + \xi_k R_{im} R'_{jl} + \xi_k R_{lm} R'_{ji} \\
& + \xi_l R_{jk} R'_{im} + \xi_l R_{ji} R'_{km} + \xi_l R_{jm} R'_{ki} \\
& + \xi_l R_{ki} R'_{jm} + \xi_l R_{km} R'_{ji} + \xi_l R_{im} R'_{jk} \\
& + \xi_m R_{jk} R'_{li} + \xi_m R_{jl} R'_{ki} + \xi_m R_{ji} R'_{kl} \\
& + \xi_m R_{kl} R'_{ji} + \xi_m R_{ki} R'_{jl} + \xi_m R_{li} R'_{jk},
\end{aligned}
$$

with i,j,k,l∈{0,...,3}, are invariant under translation by $T_1$T1. Similarly the forms $S_{ijkl}$Sijkl, defined as above but with each *R* , R′ replaced by *S* , S′, are invariant under translation by $T_2$T2. The $R_{ijk}$and $S_{ijk}$ each generate spaces of dimension 12 that intersect in a space of dimension 4. This intersection provides us with a map on the Kummer surface of $J_r$Jr.

We find a basis ξ˜0,...,ξ˜3 for ϕ(OJ˜r(2ΘJ˜r)) that is the pullback of a basis of the type described by (2). We can then read off the curve $\tilde{C}_r$ C˜r, at least up to quadratic twist, from the resulting equation for K˜r. The basis choice can largely be characterized by the order of vanishing of each $\xi_i$ξi at the identity element. This leads us to conclude that, up to scalar multiples, we should take the basis choice

$$
\begin{aligned}
\tilde{\xi}_0 &= (1\xi_0 + 0\xi_1 + 0\xi_2)\xi_3^4 + \cdots, \\
\tilde{\xi}_1 &= (0\xi_0 + 1\xi_1 + 0\xi_2)\xi_3^4 + \cdots, \\
\tilde{\xi}_2 &= (0\xi_0 + 0\xi_1 + 1\xi_2)\xi_3^4 + \cdots .
\end{aligned}
$$

The determination of ξ˜3 is a little more involved. The resulting quintic forms are too large to reproduce here, but we have made them available electronically at [5]. We can then find the quartic satisfied by these quintic forms, and read the equation of C˜r, up to a quadratic twist, from its Kummer surface equation. We can then find the correct twist by applying the map to a member of $J_r(Q)$Jr(Q). This gives the following result, shown in the Maple file at [5].

**Theorem 4.** *Let* $C_r$ *be as described by Lemma 3. Then* J˜r=Jr/Σ*is the Jacobian of the genus* 2 *curve*

$$\widetilde{C}_r\colon y^2 = \tilde{\lambda}_1 \widetilde{G}_1(x)^2 + \widetilde{H}_1(x)^5 = \tilde{\lambda}_2 \widetilde{G}_2(x)^2 + \widetilde{H}_2(x)^5,$$

*where* $\widetilde{G}_1(x) = x^2 + 100r^2(15+4\sqrt{5})$ $\widetilde{G}_2(x) = x^2 + 100r^2(15-4\sqrt{5})$ $\tilde{\lambda}_1 = 25r(-5+2\sqrt{5})$ $\tilde{\lambda}_2 = 25r(-5-2\sqrt{5})$ $\widetilde{H}_1(x) = x - 10r\sqrt{5}$ $\widetilde{H}_2(x) = x + 10r\sqrt{5}$*. The curve* $C_r$ *is defined over k, and there is a* (5,5)-*isogeny* $\phi : J_r \mapsto \widetilde{J}_r$ *defined over k with kernel* $\langle T_1, T_2 \rangle$ *. The dual isogeny* $\widetilde{\phi} : \widetilde{J}_r \mapsto J_r$ *kernel* $\langle \widetilde{T}_1, \widetilde{T}_2 \rangle$*, where*

$$\widetilde{T}_1 = [(10r\sqrt{5}, 2000r^2\sqrt{10r(-5+\sqrt{5})}) - \infty],$$

$$\widetilde{T}_2 = [(-10r\sqrt{5}, 2000r^2\sqrt{10r(-5-\sqrt{5})}) - \infty].$$

# 5. ISOGENY DESCENT

Galois cohomology associates with an isogeny

$$0 \to J[\phi] \to J \xrightarrow{\phi} \widetilde{J} \to 0$$

between abelian varieties over a field *k* an exact sequence

$$0 \to \widetilde{J}(k)/\phi J(k) \xrightarrow{\gamma} H^1(k, J[\phi]) \to H^1(k, J).$$

For *k* a number field and *v* a place of *k* , we consider the completion $k_v$kv and its separable closure kvsep and identify Gal(kvsep/kv) with a relevant decomposition group inside Gal($k^{sep}$/k)Gal. This allows us to consider restriction maps $res_v: H^i(k,.) \to H^i(k_v,.)$. Writing $\gamma_v$ for the relevant connecting homomorphism over the base field $k_v$, this allows us to define the *Selmer group*

$$\mathrm{Sel}^{\phi}(J/k) = \{\delta \in H^1(k, J[\phi]) : \mathrm{res}_v(\delta) \in \mathrm{im}\gamma_v \text{ for all places } v \text{ of } k\}. \quad (5)$$

The Selmer group contains the image of $\gamma$ . If this containment is strict then part of the Selmer group represents non-trivial elements in Ш(J/k). To be precise, we have

$$0 \to \tilde{J}(k)/\phi J(k) \to \mathrm{Sel}^{\phi}(J/k) \to \text{Ш}(J/k)[\phi] \to 0.$$

Therefore, the computation of Selmer groups can be used to exhibit non-trivial elements in Tate–Shafarevich groups.

**Lemma 5.**

*Let* $C_r$ *and* $\tilde{C}_r$ *be as in Lemma 3and Theorem 4, with Jacobians* $J_r$, $\tilde{J}_r$, *and* (5,5)-*isogenies* $\phi: J_r \to \tilde{J}_r$, *Then there is a homomorphism*

$$\begin{aligned} q\colon \tilde{J}_r(k)/\phi J_r(k) &\longmapsto ((k[t]/(t^4+10rt^2+5r^2))^*/((k[t]/(t^4+10rt^2+5r^2))^*)^5)^{\times 2} \\ &= k\Big(\sqrt{r(-5+2\sqrt{5})}\Big)^*/\Big(k\Big(\sqrt{r(-5+2\sqrt{5})}\Big)^*\Big)^5 \\ &\quad \times k\Big(\sqrt{r(-5-2\sqrt{5})}\Big)^*/\Big(k\Big(\sqrt{r(-5-2\sqrt{5})}\Big)^*\Big)^5 \end{aligned}$$

*given by* $[P_1 + P_2 - 2\infty] \mapsto \gamma(P_1)\gamma(P_2)$, *where*

$$\begin{aligned} \gamma : (x,y) \longmapsto [&y - 5\sqrt{r(-5+2\sqrt{5})}\,(x^2 + (15+4\sqrt{5})100r^2), \\ &y - 5\sqrt{r(-5-2\sqrt{5})}\,(x^2 + (15-4\sqrt{5})100r^2)]. \end{aligned}$$

*Similarly*

$$\tilde{q}\colon J_r(k)/\tilde{\phi}\tilde{J}_r(k) \longmapsto (k[t]/(t^2-5r))^*/((k[t]/(t^2-5r))^*)^5$$
$$\times(k[t]/(t^2-r))^*/((k[t]/(t^2-r))^*)^5$$
$$= k(\sqrt{5r})^*/(k(\sqrt{5r})^*)^5 \times k(\sqrt{r})^*/(k(\sqrt{r})^*)^5$$

*is given by* $[P_1+P_2-2\infty] \mapsto \gamma(P_1)\gamma(P_2),$ *where*

$$\tilde{\gamma} : (x,y) \longmapsto [y-\sqrt{5r}(3x^2-12rx+16r^2),$$
$$y-\sqrt{r}(5x^2-20rx+16r^2)].$$

*The image of q is in the kernel of the norm map to* $k(\sqrt{5})^*/(k(\sqrt{5})^*)^5 \times k(\sqrt{5})^*/(k(\sqrt{5})^*)^5$. *The image of* $\tilde{q}$ q˜*is in the kernel of the norm map* $k^*/(k^*)^5 \times k^*/(k^*)^5$. *For any completion* $k_v$kv, *we let*

$$j_v : \left((k[t]/(t^4+10rt^2+5r^2))^*/((k[t]/(t^4+10rt^2+5r^2))^*)^5\right)^{\times 2}$$
$$\longrightarrow \left((k_v[t]/(t^4+10rt^2+5r^2))^*/((k_v[t]/(t^4+10rt^2+5r^2))^*)^5\right)^{\times 2}$$

*and*

$$\tilde{j}_v : (k[t]/(t^2-5r))^*/((k[t]/(t^2-5r))^*)^5 \times (k[t]/(t^2-r))^*/((k[t]/(t^2-r))^*)^5$$
$$\longrightarrow (k_v[t]/(t^2-5r))^*/((k_v[t]/(t^2-5r))^*)^5 \times (k_v[t]/(t^2-r))^*/((k_v[t]/(t^2-r))^*)^5$$

*be the natural maps. Let* $q_v$ *be the map on* $\tilde{J}_r(k_v)/\phi(J_r(k_v))$J*and* q˜v*the map on* $J_r(k_v)/\tilde{\phi}(\tilde{J}_r(k_v))$ *the local versions of q,* q˜. *Then the Selmer groups can be described explicitly as the intersections over v of* $j_v^{-1}(q_v(\tilde{J}_r(k_v)/\phi(J_r(k_v))))$ *and* $\tilde{j}_v^{-1}(\tilde{q}_v(J_r(k_v)/\tilde{\phi}(\tilde{J}_r(k_v))))$

**Proof.** This is a direct application of the theory developed in [3] and [6].

We take *S* to be the set of primes consisting of 5 and the primes of bad reduction of $C_r$. By [3, Proposition 9.2], the Selmer groups lie in the subgroups that are unramified outside *S*. We can represent those using *S*-units in the above number fields. Note that, for any curve in our family, the descent can be performed using at worst quartic number fields. This already provides us with explicit finite groups that contain the Selmer groups. The remaining conditions come from the local images at v∈S (note that $R^{\times}/R^{\times 5}$ is trivial, so the archimedean place does not provide any information). With the explicit description of the maps $q_v$qv and $\tilde{q}_v$ we can generate elements in their images. Suppose now that k=Q. Using [6, Lemma 3.8, Proposition 3.9] we have

$$\#\tilde{J}_r(\mathbb{Q}_v)/\phi(J_r(\mathbb{Q}_v))\,\#J_r(\mathbb{Q}_v)/\tilde{\phi}(\tilde{J}_r(\mathbb{Q}_v)) = \frac{\#J_r[\phi](\mathbb{Q}_v)\,\#\tilde{J}_r[\tilde{\phi}](\mathbb{Q}_v)}{|5|_p^2}. \tag{6}$$

So we know when we have found enough elements to generate the entire image. By explicitly computing the restrictions $j_v$jv and $\tilde{j}_v$j˜v, we can compute the Selmer groups using essentially the definition in (5).

We now give two examples. The first example will illustrate a situation where a (5,5)(5,5)-isogeny descent gives a rank bound of 0; this is the same as the bound obtained by a complete 2-descent, but it only requires easier computations over fields of smaller degree. The second example has a nonzero rank bound for the (5,5)(5,5)-isogeny descent (which can be shown, using only the easier direction of the isogeny for which only a quadratic number field is required for the Cassels map), and a rank bound of 0 for complete 2-descent, giving a nontrivial element in Ш(J/Q)[5]Ш(J/Q)[5]. The computations involved in these examples are given in detail at the end of the Maple file at [5].

**Example 6.** Let $J_1$J1 be the Jacobian of the curve

$$C_1 : y^2 = 5(3x^2 - 12x + 16)^2 + (x-4)^5 = (5x^2 - 20x + 16)^2 + x^5.$$

Then $J_1(Q)$ has a rank bound of 0 using descent via (5,5)-isogeny.

**Proof.**

For r=1r=1, we have

$$\widetilde{C}_1 : y^2 = 25(-5+2\sqrt{5})(x^2+100(15+4\sqrt{5}))^2+(x-10\sqrt{5})^5$$
$$= 25(-5-2\sqrt{5})(x^2+100(15-4\sqrt{5}))^2+(x+10\sqrt{5})^5,$$

also defined over Q. We note here that the quartic number field can be identified with Q($\zeta_5$), and so we shall express our maps in terms of elements of this field. The kernel of $\phi : J_1 \to \widetilde{J}_1$ is generated by

$$T_1 = [(4, 16\sqrt{5}) - \infty],$$
$$T_2 = [(0, 16) - \infty].$$

The kernel of $\tilde{\phi} : \widetilde{J}_1 \to J_1$ is generated by

$$\widetilde{T}_1 = [(10(1+2\zeta_5+2\zeta_5^4), 4000(2+4\zeta_5+\zeta_5^2+3\zeta_5^3)) - \infty]$$
$$\widetilde{T}_2 = [(10(-1-2\zeta_5-2\zeta_5^4), 4000(1+2\zeta_5+3\zeta_5^2-\zeta_5^3) - \infty]$$

Then, for r=1r=1,

$$q\colon \widetilde{J}_1(k)/\phi J_1(k) \longmapsto \mathbb{Q}(\zeta_5)^*/(\mathbb{Q}(\zeta_5)^*)^5 \times \mathbb{Q}(\zeta_5)^*/(\mathbb{Q}(\zeta_5)^*)^5$$

is given by $[P_1+P_2-2\infty] \mapsto \gamma(P_1)\gamma(P_2)$, where

$$\gamma : (x,y) \longmapsto [y-5(1+2\zeta_5+2\zeta_5^3)(x^2+100(15+4(1+2\zeta_5+2\zeta_5^4))),$$
$$y-5(1+2\zeta_5+2\zeta_5^2)(x^2+100(15-4(1+2\zeta_5+2\zeta_5^4)))].$$

Similarly

$$\tilde{q}\colon J_1(k)/\tilde{\phi}\widetilde{J}_1(k) \longmapsto \mathbb{Q}(\sqrt{5})^*/(\mathbb{Q}(\sqrt{5})^*)^5 \times \mathbb{Q}^*/(\mathbb{Q}^*)^5$$

is given by $[P_1+P_2-2\infty] \mapsto \gamma(P_1)\gamma(P_2),$ where

$$\tilde{\gamma} : (x, y) \longmapsto [y - \sqrt{5}(3x^2 - 12x + 16),$$
$$y - (5x^2 - 20x + 16)].$$

The only bad primes are 2, 5, ∞. The units of Q($\zeta_5$) are generated by $-\zeta_5$−ζ5, 1+$\zeta_5$1+ζ5, and −1 is trivial modulo fifth powers, so we can replace $-\zeta_5$−ζ5 with $\zeta_5$ζ5. The primes above 2, 5 are 2, 1−$\zeta_5$1−ζ5, so

$$\mathrm{im}(q) \subseteq \langle [\zeta_5, 1], [1 + \zeta_5, 1], [2, 1], [1 - \zeta_5, 1], [1, \zeta_5], [1, 1 + \zeta_5], [1, 2], [1, 1 - \zeta_5] \rangle.$$

Note that the norms from Q($\zeta_5$) to $\mathbb{Q}(\zeta_5)$ to $\mathbb{Q}(\sqrt{5})$ of $\zeta_5$, $1 + \zeta_5$, 2, $1 - \zeta_5$ are 1, $(3 + \sqrt{5})/2 (5 - \sqrt{5})/2$After taking into account that the image of $q$ is contained in the kernel of the norm map to $\mathbb{Q}(\sqrt{5})^*/(\mathbb{Q}(\sqrt{5})^*)^5$, we can reduce to ⟨[$\zeta_5$,1],[1,$\zeta_5$]⟩ ⟩ . After taking into account that the second component of $q$ is the image of the first component under $\zeta_5 \mapsto \zeta_5^2$, we see that in fact

$$\mathrm{im}(q) \subseteq \langle [\zeta_5, \zeta_5^2] \rangle.$$

The units of $\mathbb{Q}(\sqrt{5})$ are generated by $(1 - \sqrt{5})/2$and the primes above 2, 5 are 2, $\sqrt{5}$, so the first component of $\tilde{q}$ is contained in $\langle (1 - \sqrt{5})/2, 2, \sqrt{5} \rangle$. These have norms −1, 4, −5, so that in fact the first component is contained in $\langle (1 - \sqrt{5})/2 \rangle$. The second component of $\tilde{q}$ is contained in ⟨2,5. Hence

$$\mathrm{im}(\tilde{q}) \subseteq \langle [(1 + \sqrt{5})/2, 1], [1, 2], [1, 5] \rangle.$$

Taking into account that #$J_1$[5](Q)=5#, we already have an a priori bound of 3 on the rank.

First consider p=2. Using Eq. (6) and the fact that $\#J_1[\phi](\mathbb{Q}_2) = 5^1$, and $\#\tilde{J}_1[\tilde{\phi}](\mathbb{Q}_2) = 5^0$, we have

$$\#\tilde{J}_1(\mathbb{Q}_2)/\phi(J_1(\mathbb{Q}_2)) \, \#J_1(\mathbb{Q}_2)/\tilde{\phi}(\tilde{J}_1(\mathbb{Q}_2)) = 5^1.$$

Note that substituting x=−2x=−2 into the quintic of $\widetilde{C}_1$ gives $-2^8 \cdot 330\,047$, which is a square in $Q_2$. Let $y_0 \in Q_2$y0∈Q2 be such that $y_0^2 = -2^8 \cdot 330\,047$ with $y_0 \equiv 48 \pmod{64}$ Then $[(-2, y_0) - \infty] \mapsto [\zeta_5, \zeta_5^2]$ and $[\zeta_5, \zeta_5^2] \neq [1,1]$ in $\mathbb{Q}_2(\zeta_5)^*/(\mathbb{Q}_2(\zeta_5)^*)^5 \times \mathbb{Q}_2(\zeta_5)^*/(\mathbb{Q}_2(\zeta_5)^*)^5$. Hence we have now found all $\#J_1(\mathbb{Q}_2)/\phi(J_1(\mathbb{Q}_2)) = 5^1$ and $\#J_1(\mathbb{Q}_2)/\phi(J_1(\mathbb{Q}_2)) = 5^0$. Also, $j_2$ has trivial kernel and j˜2 has kernel $\langle [(1-\sqrt{5})/2, 1], [1,5] \rangle$. Hence

$$j_2^{-1}(q_2(\widetilde{J}_1(\mathbb{Q}_2)/\phi(J_1(Q_2)))) = \langle [\zeta_5, \zeta_5^2] \rangle,$$
$$\tilde{j}_2^{-1}(\tilde{q}_2(J_1(\mathbb{Q}_2)/\tilde{\phi}(\widetilde{J}_1(Q_2)))) = \langle [(1-\sqrt{5})/2, 1], [1,5] \rangle.$$

At this point, the bound of the rank has been lowered to 2.

We now consider p=5. Using Eq. (6) and the fact that $\#J_1[\phi](\mathbb{Q}_2) = 5^1$, and $\#\widetilde{J}_1[\tilde{\phi}](\mathbb{Q}_2) = 5^0$, we have

$$\#\widetilde{J}_1(\mathbb{Q}_5)/\phi(J_1(\mathbb{Q}_5))\,\#J_1(\mathbb{Q}_5)/\tilde{\phi}(\widetilde{J}_1(\mathbb{Q}_5)) = 5^3.$$

Note that substituting x=1x=1 into the quintic of $\widetilde{C}_1$ gives a square in $Q_5$, so let $y_1 \in Q_5$ be such that $(1, y_1) \in \widetilde{C}_1(\mathbb{Q}_5)$Then $q_5$ maps $[(1,y_1)-\infty]$ to a nontrivial element which is not even in the image of $j_5$. On $C_1$, note that there is a point $(3,y_2) \in C_1(Q_5)$ and that $\tilde{q}_5 : [(1, y_1) - \infty] \mapsto [1,2]$ Also $\tilde{q}_5 : T2 \mapsto [(1-\sqrt{5})/2, 1]$ Furthermore, [1,2] and $[(1-\sqrt{5})/2, 1]$ [(1−5)/2,1] are independent in $\mathbb{Q}_5(\sqrt{5})^*/(\mathbb{Q}_5(\sqrt{5})^*)^5 \times \mathbb{Q}^*/(\mathbb{Q}^*)^5$ We have now found all of $\#\widetilde{J}_1(\mathbb{Q}_5)/\phi(J_1(\mathbb{Q}_5)) = 5^1$# and $\#J_1(\mathbb{Q}_5)/\tilde{\phi}(\widetilde{J}_1(\mathbb{Q}_5)) = 5^2$Also, $j_5$ and $\tilde{j}_5$ have trivial kernel. Hence

$$j_5^{-1}(q_5(\widetilde{J}_1(\mathbb{Q}_5)/\phi(J_1(Q_5)))) = \langle [1,1] \rangle,$$
$$\tilde{j}_5^{-1}(\tilde{q}_5(J_1(\mathbb{Q}_5)/\tilde{\phi}(\widetilde{J}_1(Q_5)))) = \langle [(1-\sqrt{5})/2, 1], [1,2] \rangle.$$

Taking the intersection of the information at p=2 and p=5, we see that im(q) is the trivial group, and so $\widetilde{J}_1(\mathbb{Q})/\phi(J_1(Q))$ is the trivial group, and $\mathrm{im}(\tilde{q})$is contained in $\langle[(1-\sqrt{5})/2,1]\rangle$, and indeed these are equal, since $\tilde{q}:T_2\mapsto[(1-\sqrt{5})/2,1]$ We finally deduce that $\#J_1(\mathbb{Q})/5J_1(\mathbb{Q})=5^1$, and $\#J_1(\mathbb{Q})[5]=5^1$, so that the rank is 0. □

The same result can also be obtained by performing a 2-descent, for example using Magma [1], which requires working over a degree 5 number field.

For the next example, we do not find completely the Selmer bound for the (5,5)-isogeny, but we do enough to show that the bound is nonzero, and deduce the existence of 5-part of Ш by comparison with a complete 2-descent, for which the rank bound is 0. We first require the following technical lemma.

**Lemma 7.**

*Let p be prime, satisfying* $p\equiv 1$ *or* $4\ (mod\ 5)$*and* $p\equiv 3\ (mod\ 4)$ *and let* $\epsilon_1=a+b\sqrt{5p}$*, with* $a,b\in\mathbb{Z}$ *and* $a,b>0$ *, be a fundamental unit of* $\mathbb{Z}[\sqrt{5p}]$ *Let* $w_0\in\mathbb{Q}_5$ *satisfy* $w_0^2=p$ w02=p*(such* $w_0$ *exists, since* $p\equiv 1$*or* $4\ (mod\ 5)$*). Suppose that* $(w_0^5-w_0^4\sqrt{5p})/2=\epsilon_1^i$ $\mathbb{Q}_5(\sqrt{5p})^*/(\mathbb{Q}_5(\sqrt{5p})^*)^5$ *for some* $i=1,\ldots 4$*, that a is odd, b is even, and that a is a quadratic residue mod p. Then the rank bound on* $J_p(\mathbb{Q})$ *from descent via* (5,5)*-isogeny is nonzero.*

**Proof.** The bad primes here are: 2, 5, *p* , ∞. Consider whether the element $[\epsilon_1,1]$ can be in the image of $\tilde{q}$.

First note that $[\epsilon_1,1]$ is in the kernel of $\tilde{j}_\infty$ and so $[\epsilon_1,1]$ cannot be disqualified by local arguments at infinity.

Let $f(x)=x^5-\epsilon_1$, considered over $\mathbb{Q}_2(\sqrt{5p})$. Then, since *a* is odd and *b* is even, we have $|f(1)|_5<1|$ and $|f'(1)|_2=1|$ and so f(x) has a root in $\mathbb{Q}_2(\sqrt{5p})$ by Hensel's Lemma. This means that $\epsilon_1\in(\mathbb{Q}_2(\sqrt{5p})^*)^5$, so that $[\epsilon_1,1]$ is in the kernel of $\tilde{j}_2$ and cannot be disqualified by local arguments at 2.

Now, let $g(x)=x^5-\epsilon_1$, considered over $\mathbb{Q}_p(\sqrt{5p})$. Since *a* is a quadratic residue mod *p* , there exists $v_0$ such that $v_0^2 \equiv a \pmod p$. Then $|g(v_0)|_p<1|$ and $|g'(v_0)|_p=1$, and so g(x) has a root in $\mathbb{Q}_p(\sqrt{5p})$ by Hensel's Lemma. This means that $\epsilon_1 \in (\mathbb{Q}_p(\sqrt{5p})^*)^5$, so that $[\epsilon_1,1]$[ϵ1,1] is in the kernel of $\tilde{j}_p$ and cannot be disqualified by local arguments at *p*.

Since we are given the existence of $w_0 \in Q_5$ satisfying $w_0^2 = p$, we now note that $T_2 = [(0, 16p^2\sqrt{p}) - \infty] \in J_p(\mathbb{Q}(\sqrt{p}))$ has an image in $J_p(Q_5)$ namely $[(0,16p^2w_0)-\infty] \in J_p(Q_5)$, which is mapped by $\tilde{q}_p$ to $[(w^5 - w^4\sqrt{5p})/2, 1]$, which we are given to be the same as $[\epsilon_1^i, 1]$ [ϵ1i,1], for some i=1,...,4. Hence again, $[\epsilon_1,1]$ cannot be disqualified by local arguments at 5.

Since also $J_p[\phi](Q)$ and $\tilde{J}_p[\tilde{\phi}](\mathbb{Q})$) are the trivial groups, it follows that the rank bound from descent via (5,5)-isogeny is nonzero. □

**Example 8.** Let $J_{11}$J11 be the Jacobian of the curve

$$\begin{aligned}\mathcal{C}_{11} : y^2 &= 5 \cdot 11(3x^2 - 12 \cdot 11x + 16 \cdot 11^2)^2 + (x - 4 \cdot 11)^5 \\ &= 11(5x^2 - 20 \cdot 11x + 16 \cdot 11^2)^2 + x^5.\end{aligned}$$

Then $J_{11}(Q)$ has a nonzero rank bound using descent via (5,5)-isogeny, and a rank bound of 0 via complete 2-descent, and so there is a nontrivial element of Ш$(J_{11}/Q)[5]$.

**Proof.** Here p=11 and the fundamental unit for $\mathbb{Q}(\sqrt{55})$ is $\epsilon_1 = a + b\sqrt{55} = 89 + 12\sqrt{55}$, which satisfies: *a* odd, *b* even, and a=89 is a quadratic residue mod 11. Let $w_0 \in Q_5$ be such that $w_0^2 = 11$, and choose $w_0 \equiv 1 \pmod 5$ Note that then $w_0 \equiv 56 \pmod{5^3}$). Consider $h(x) = x^5 - \epsilon_1^2(w_0^5 - w_0^4\sqrt{55})$ Then $|h(56 \cdot 18 + 15\sqrt{55})|_5 \le 5^{-3}|$ (as can be seen, just using the approximation 56 for $w_0$w0) and $|h'(56 \cdot 18 + 15\sqrt{55})|_5 = 5^{-1}|$, so by Hensel's Lemma, $\epsilon_1^2(w_0^5 - w_0^4\sqrt{55}) \in (\mathbb{Q}_5(\sqrt{55})^*)^5$, giving that $w_0^5 - w_0^4\sqrt{55} = \epsilon_1^3$ in $\mathbb{Q}_5(\sqrt{55})^*/(\mathbb{Q}_5(\sqrt{55})^*)^5$, as required. Hence by the previous lemma,

$[\epsilon_1,1]$ represents a nontrivial element in the Selmer group, and the rank bound on $J_{11}(Q)$ from descent via (5,5)-isogeny is nonzero.

On the other hand, performing a descent via 2-isogeny one obtains a rank bound of 0, so that there must exist a nontrivial element in Ш($J_{11}$/Q)[5]Ш(J11/Q)[5]. □

## REFERENCES


1. The MAGMA computer algebra system is described in Wieb Bosma, John Cannon, Catherine Playoust, The Magma algebra system. I. The user language, J. Symbolic Comput. 24 (3–4) (1997) 235–265.

2. Nils Bruin, E. Victor Flynn, Damiano Testa, Descent via (3, 3)-isogeny on Jacobians of genus 2 curves, Acta Arith. 165 (2014) 201–223.

3. Nils Bruin, Bjorn Poonen, Michael Stoll, Generalized explicit descent and its application to curves of genus 3, available at arXiv:1205.4456, 2012.

4. J.W.S. Cassels, E.V. Flynn, Prolegomena to a Middlebrow Arithmetic of Curves of Genus 2, London Mathematical Society Lecture Note Series, vol. 230, Cambridge University Press, Cambridge, 1996, MR1406090 (97i:11071).

5. E. Victor Flynn, Electronic resources, http://people.ox.ac.uk/flynn/genus2/pn3, 2014. [6] Edward F. Schaefer, Computing a Selmer group of a Jacobian using functions on the curve, Math. Ann. 310 (3) (1998) 447–471, MR1612262 (99h:11063).

6. Michael Stoll, Two simple 2-dimensional abelian varieties defined over Q with Mordell–Weil rank at least 19, C. R. Acad. Sci. Paris Sér. I Math. 321 (1995) 1341–1344.

7

# NEW EXTENSIONS OF SOME KNOWN SPECIAL POLYNOMIALS UNDER THE THEORY OF MULTIPLE Q-CALCULUS

**Mehmet Acikgoz[1], Serkan Araci[2], Uğur Duran[1]**

[1]Department of Mathematics, Faculty of Arts and Science, University of Gaziantep, Gaziantep, Turkey

[2]Department of Economics, Faculty of Economics, Administrative and Social Science, Hasan Kalyoncu University, Gaziantep, Turkey

## ABSTRACT

In the year 2014, the idea of multiple $q$-calculus was formulated and introduced in the book of Nalci and Pashaev [9] in which this idea is simple but elegant method in order to derive new generating functions of some special polynomials that are generalizations of known $q$-polynomials. In this paper, we will use Nalci and Pashaev's method in order to find a systematic study of new types of the Bernoulli polynomials, Euler polynomials and Genocchi polynomials. Also we will obtain recursive formulas for these polynomials.

## KEYWORDS

Quantum calculus, Multiple quantum calculus, q-Bernoulli polynomials, q-Euler polynomials, q-Genocchi polynomials, Generating function

# 1. INTRODUCTION

## 1.1. $q$-Calculus

The usual quantum calculus (or recalled $q$-calculus) has been extensively studied for a long time by many mathematicians, physicists and engineers. The development of $q$-calculus stems from the applications in many fields such as engineering, economics, math-ematics, and so on. One of the important branches of $q$-calculus is $q$-special polynomials. For example, Kim [18] constructed$q$-generalized Euler polynomials based on $q$-exponential function. Moreover, Srivastava *et al* investigated Apostol $q$-Bernoulli, Apostol $q$-Euler polynomials and Apostol $q$-Genocchi polynomials. This is why $q$-calculus is thought as one of the useful tools to study with special numbers and polynomias. For more information related these issues, see, e.g. [1,2,3,5,6,8-13,16-21].

Before starting at multiple $q$-calculus, we first give some basic notations about $q$-calculus which can be found in [3].

For a real number (or complex number) $x$, $q$-number (quantum number) is known as

$$[x]_q := \begin{cases} \frac{1-q^x}{1-q}, & \text{if } q \neq 1, \\ x, & \text{if } q = 1 \end{cases} \tag{1.1}$$

which is also called non-symmetrical $q$-number. The followings can be easily derived using (1.1):

$$[x+y]_q = [x]_q + q^x [y]_q \quad (q\text{-addition formula}) \tag{1.2}$$

$$[x-y]_q = -q^{x-y}[y]_q + [x]_q \quad (q\text{-substraction formula}) \tag{1.3}$$

$$[xy]_q = [x]_q [y]_{q^x} \quad (q\text{-product rule}) \tag{1.4}$$

$$\left[\frac{x}{y}\right]_q = \frac{[x]_q}{[y]_{q^{\frac{x}{y}}}} = \frac{[x]_{q^{\frac{1}{y}}}}{[y]_{q^{\frac{1}{y}}}} \quad (q\text{-division rule}) \tag{1.5}$$

where $x, y$ are real or complex numbers.

The $q$-binomial coefficients are defined for positive integer $n, k$ as

$$\begin{bmatrix} n \\ k \end{bmatrix}_q = \frac{[n]_q!}{[k]_q![n-k]_q!} \tag{1.6}$$

where $[n]_q! = [n]_q[n-1]_q[n-2]_q \cdots [1]_q$, $n = 1, 2, \ldots$; $[0]_q! = 1.$

The $q$-derivative $D_q f(x)$ of a function $f$ is given as

$$D_q f(x) = \frac{f(x) - f(qx)}{(1-q)x}, q \neq 1, x \neq 0, D_q f(0) = f'(0),$$

provided $f'(0)$ exists.

For any $z \in \mathbb{C}$ with $|z| < 1$,

$$e_q(z) = \sum_{n=0}^{\infty} \frac{z^n}{[n]_q!} \text{ and } E_q(z) = \sum_{n=0}^{\infty} q^{\binom{n}{2}} \frac{z^n}{[n]_q!}.$$

For the $q$-commuting variables $x$ and $y$ such as $yx = qxy$, we know that

$$e_q(x+y)=e_q(x)e_q(y).$$

The $q$-integral was defined by Jackson as follows:

$$\int_0^x f(y)d_q y=(1-q)x\sum_{n=0}^{\infty} f\left(q^n x\right)q^n$$

provided that the series on right hand side converges absolutely.

## 1.2. Multiple $q$-calculus

All notations and all corollaries written in this part have been taken from the Book of Nalci and Pashaev [9].

Consider basis vector $\vec{q}$ with coordinates $q_1, q_2, \ldots, q_N$ so that the multiple $q$-number can be de.ned as

$$[n]_{q_i,q_j}:=\frac{q_i^n-q_j^n}{q_i-q_j}=[n]_{q_j,q_i}, \tag{1.7}$$

which is symmetric. Hence, we can write $N\times N$ matrix with $q$-numbers elements in the following form:

$$\left([n]_{q_i,q_j}\right)=\begin{pmatrix} [n]_{q_1,q_1} & [n]_{q_1,q_2} & \cdots & [n]_{q_1,q_N} \\ [n]_{q_2,q_1} & [n]_{q_2,q_2} & & [n]_{q_2,q_N} \\ \cdots & \cdots & \cdots & \cdots \\ [n]_{q_N,q_1} & [n]_{q_N,q_2} & \cdots & [n]_{q_N,q_N} \end{pmatrix}.$$

Diagonal terms of this matrix are defined in the limit $q_j\to q_i$ as

$$\lim_{q_j\to q_i}[n]_{q_i,q_j}=\lim_{q_j\to q_i}\frac{q_i^n-q_j^n}{q_i-q_j}=nq_i^{n-1}. \tag{1.8}$$

So, by (1.8), we see that this symmetric matrix can be shown as

$$\left([n]_{q_i,q_j}\right)=\begin{pmatrix} nq_1^{n-1} & [n]_{q_1,q_2} & \cdots & [n]_{q_1,q_N} \\ [n]_{q_2,q_1} & nq_2^{n-1} & & [n]_{q_2,q_n} \\ \cdots & \cdots & \cdots & \cdots \\ [n]_{q_N,q_1} & [n]_{q_N,q_2} & \cdots & nq_N^{n-1} \end{pmatrix}.$$

The followings can be easily derived using (1.7):

$$[n+m]_{q_i,q_j}=q_i^n[m]_{q_i,q_j}+q_j^m[n]_{q_i,q_j}$$

($q$-multiple addition formula)

$$[n-m]_{q_i,q_j}=-q_j^{-m}\left([n]_{q_i,q_j}-q_i^{n-m}[m]_{q_i,q_j}\right)$$

($q$-multiple substraction formula)

$$[nm]_{q_i,q_j}=[m]_{q_i,q_j}[n]_{q_i^m,q_j^m}$$

($q$-multiple product rule)

$$\left[\frac{n}{m}\right]_{q_i,q_j}=\frac{[n]_{q_i,q_j}}{[m]_{q_i^{\frac{n}{m}},q_j^{\frac{n}{m}}}}=\frac{[n]_{q_i^{\frac{1}{m}},q_j^{\frac{1}{m}}}}{[m]_{q_i^{\frac{1}{m}},q_j^{\frac{1}{m}}}}$$

($q$-multiple division rule)

where $n$, $m$ are real or complex numbers.

In multiple $q$-calculus, multiple $q$-derivative with base $q_i$, $q_j$ is given by

$$D_{q_i,q_j}f(x)=\frac{f(q_ix)-f(q_jx)}{(q_i-q_j)x}$$

representing $N \times N$ matrix of multiple $q$-derivative operators $D := \left(D_{q_i,q_j}\right)$ which is sym-metric: $D_{q_i,q_j} = D_{q_j,q_i}$ where *i* and $j = 1,2,\ldots,N.$

**Corollary 1.** *For N = 1 case and* $q_1 = q_2 \equiv q,$ *we have*

$$D = \left(D_{q_i,q_j}\right) = \begin{pmatrix} D_{q_1,q_1} & D_{q_1,q_2} & \cdots & D_{q_1,q_N} \\ D_{q_2,q_1} & D_{q_2,q_2} & \cdots & D_{q_2,q_N} \\ \cdots & \cdots & \cdots & \cdots \\ D_{q_N,q_1} & D_{q_N,q_2} & \cdots & D_{q_N,q_N} \end{pmatrix}.$$

$$[n]_{q,q} = nq^{n-1} \ \ and \ \ D_{q,q} = M_q \frac{d}{dx}$$

where $M_q = q^{x\frac{d}{dx}}.$ Also, in the case $q = 1,$ we have the standard number $[n]_{1,1} = n$ and the usual derivative $D_{1,1} = \frac{d}{dx}.$

**Corollary 2.** *For N = 2 case, we have*

$$[n]_{q_1,q_1} = nq_1^{n-1}, [n]_{q_1,q_2} = [n]_{q_2,q_1} = \frac{q_1^n - q_2^x}{q_1 - q_2},$$

$$[n]_{q_2,q_2} = nq_2^{n-1}$$

$$D_{q_1,q_1} = M_{q_1} \frac{d}{dx}, D_{q_1,q_2} = D_{q_2,q_1} = \frac{M_{q_1} - M_{q_2}}{(q_1 - q_2)x},$$

$$D_{q_2,q_2} = M_{q_2} \frac{d}{dx}$$

**Corollary 3.** *Choosing* $q_1$ = 1 *and* $q_2$ = *q gives non-symmetrical case as*

$$[n]_{1,1} = n, \ [n]_{1,q} = [n]_{q,1} = [n]_q, [n]_{q,q} = nq^{n-1},$$

$$D_{1,1} = \frac{d}{dx}, D_{1,q} = D_{q,1} = \frac{1 - M_q}{(1-q)^x}, D_{q,q} = M_q \frac{d}{dx}.$$

**Corollary 4.** Taking $q_1 = q$ and $q_2 = \frac{1}{q}$ gives symmetrical case as

$$[n]_{q,q} = nq^{n-1},\ [n]_{1,\frac{1}{q}} = [n]_{\tilde{q}}, [n]_{\frac{1}{q},\frac{1}{q}} = n\left(\frac{1}{q}\right)^{n-1},$$

$$D_{q,q} = M_q \frac{d}{dx},\ D_{\frac{1}{q},q} = D_{q,\frac{1}{q}} = \frac{M_{\frac{1}{q}} - M_q}{\left(\frac{1}{q} - q\right)^x},$$

$$D_{\frac{1}{q},\frac{1}{q}} = M_{\frac{1}{q}} \frac{d}{dx}.$$

The multiple $q$-analogue of $(x-a)^n$ is the polynomial

$$(x+a)^n_{q_i,q_j} := \begin{cases} \left[\begin{matrix} \left(x+q_i^{n-1}a\right)\left(x+q_i^{n-2}q_j a\right)\ldots \\ \left(x+q_i q_j^{n-2}a\right)\left(x+q_j^{n-2}a\right) \end{matrix}\right], & \text{if } n \geq 1 \\ 1, & \text{if } n = 1 \end{cases}$$

or equivalently

$$(x+a)^n_{q_i,q_j} = \sum_{k=0}^{n} \begin{bmatrix} n \\ k \end{bmatrix}_{q_i,q_j} (q_i q_j)^{\frac{k(k-1)}{2}} x^{n-k} a^k$$

where $x$ and a is commutative, $xa = ax$. $q$-multiple Binomial coefficients and multiple $q$-factorial are defined by

Two types of multiple $q$-exponential functions are de.ned by

$$[n]_{q_i,q_j}! = \begin{cases} [n]_{q_i,q_j}[n-1]_{q_i,q_j}\ldots[2]_{q_i,q_j}[1]_{q_i,q_j} & \text{if } n \geq 1 \\ 1 & \text{if } n \geq 1 \end{cases} \quad (n \in \mathbb{N}).$$

$$e_{q_i,q_j}(x)=\sum_{n=0}^{\infty}\frac{x^n}{[n]_{q_i,q_j}!}$$

$$E_{q_i,q_j}(x)=\sum_{n=0}^{\infty}(q_iq_j)^{\frac{n(n-1)}{2}}\frac{x^n}{[n]_{q_i,q_j}!}$$

which satisfy the following condition for commutative $x$ and $y$, $xy = yx$

$$e_{q_i,q_j}(x+y)_{q_iq_j}=e_{q_i,q_j}(x)E_{q_i,q_j}(y).$$

The generalization of Jackson's integral (called multiple $q$-integral) is given by

$$\int f\left(\frac{x}{q_i}\right)d_{\frac{q_j}{q_i}}x=(q_i-q_j)\sum_{k=0}^{\infty}\frac{q_j^kx}{q_i^{k+1}}f\left(\frac{q_j^k}{q_i^{k+1}}x\right)$$

Let $f(x)=\sum_{k=0}^{\infty}a_kx^k$ be formal power series. Applying multiple $q$-integral to the both sides of $f(x)$ gives

$$\int f\left(\frac{x}{q_i}\right)d_{\frac{q_j}{q_i}}x=\sum_{k=0}^{\infty}q_i^{k+1}a_k\frac{x^{k+1}}{[k+1]_{q_i,q_j}}+C$$

where $C$ is constant.

In the next section, we will use Nalci and Pashaev's method in order to find a systematic study of new types of the Bernoulli polynomials, Euler polynomials and Genocchi polynomials. Also we will obtain recursive formulas for these polynomials.

## 2 MAIN RESULTS

Recently, analogues of Bernoulli, Euler and Genocchi polynomials were studied by many mathematicians [1, 2, 5, 6, 11, 12, 13, 17, 18, 19, 20, 21]. We are now ready to give the definition of generating functions, corresponding to multiple $q$-calculus, of Bernoulli type, Euler type and Genocchi type polynomials.

**Definition 1.** *Let n be positive integer, we define*

$$\mathcal{S}(x,z:q_i,q_j)=\sum_{n=0}^{\infty}\mathcal{B}_n(x:q_i,q_j)\frac{z^n}{[n]_{q_i,q_j}!}$$

$$=\frac{z}{e_{q_i,q_j}(z)-1}e_{q_i,q_j}(xz)\quad(|z|<2\pi)$$

$$\mathcal{U}(x,z:q_i,q_j)=\sum_{n=0}^{\infty}\mathcal{E}_n(x:q_i,q_j)\frac{z^n}{[n]_{q_i,q_j}!}$$

$$=\frac{[2]_{q_i,q_j}}{e_{q_i,q_j}(z)+1}e_{q_i,q_j}(xz)\quad(|z|<\pi)$$

$$\mathcal{M}(x,z:q_i,q_j)=\sum_{n=0}^{\infty}\mathcal{G}_n(x:q_i,q_j)\frac{z^n}{[n]_{q_i,q_j}!}$$

$$=\frac{[2]_{q_i,q_j}z}{e_{q_i,q_j}(z)+1}e_{q_i,q_j}(xz)\quad(|z|<\pi)$$

*where* $\mathcal{B}_n(x:q_i,q_j)$, $\mathcal{E}_n(x:q_i,q_j)$ *and* $\mathcal{G}_n(x:q_i,q_j)$ *are called, respectively, Bernoulli-type,Euler-type and Genocchi-type polynomials.*

**Corollary 5.** *Taking* $q_i = q_j = 1$ *for indexes i and j in the case N = 1 in Definition 1, we have*

$$\sum_{n=0}^{\infty}B_n(x)\frac{z^n}{n!}=\frac{z}{e^z-1}e^{xz}\quad(|z|<2\pi)$$

$$\sum_{n=0}^{\infty}E_n(x)\frac{z^n}{n!}=\frac{2}{e^z+1}e^{xz}\quad(|z|<\pi)$$

$$\sum_{n=0}^{\infty}G_n(x)\frac{z^n}{n!}=\frac{2z}{e^z+1}e^{xz}\quad(|z|<\pi)$$

*where* $B_n(x)$*,* $E_n(x)$ *and* $G_n(x)$ *are called Bernoulli polynomials, Euler polynomials and Genocchi polynomials, respectively* (*see* [4, 7, 14, 15, 19]).

**Corollary 6.** *As a special case of Definition 1, we have*

$$\sum_{n=0}^{\infty} B_n(x|q)\frac{z^n}{[n]_q!} = \frac{z}{e_q(z)-1}e_q(xz) \ (|z|<2\pi)$$
$$\sum_{n=0}^{\infty} E_n(x|q)\frac{z^n}{[n]_q!} = \frac{[2]_q}{e_q(z)+1}e_q(xz) \ (|z|<\pi)$$
$$\sum_{n=0}^{\infty} G_n(x|q)\frac{z^n}{[n]_q!} = \frac{[2]_q z}{e_q(z)+1}e_q(xz) \ (|z|<\pi)$$

*where* $B_n(x|q)$, $E_n(x|q)$ *and* $G_n(x|q)$ *are called q-Bernoulli polynomials, q-Euler polynomials and q-Genocchi polynomials, respectively (see* [18, 20, 21]).

Taking $x = 0$ in the above definition, we have

$$\mathcal{B}_n(0:q_i,q_j) := \mathcal{B}_n(q_i,q_j) \quad \text{(Bernoulli-type number)}$$

$$\mathcal{E}_n(0:q_i,q_j) := \mathcal{E}_n(q_i,q_j) \quad \text{(Euler-type number)}$$
$$\mathcal{G}_n(0:q_i,q_j) := \mathcal{G}_n(q_i,q_j) \quad \text{(Genocchi-type number)}$$

and from the above, we write

$$\mathcal{S}(0,z:q_i,q_j) := \mathcal{S}(z:q_i,q_j),$$
$$\mathcal{U}(0,z:q_i,q_j) := \mathcal{U}(z:q_i,q_j),$$
$$\mathcal{M}(0,z:q_i,q_j) := \mathcal{M}(z:q_i,q_j). \tag{2.1}$$

From Definition 1 and (2.1), we get the following corollary.

**Corollary 7.** *The following functional equations hold true:*

$$\mathcal{S}(x,z:q_i,q_j) := \mathcal{S}(z:q_i,q_j)e_{q_i,q_j}(xz),$$
$$\mathcal{U}(x,z:q_i,q_j) := \mathcal{U}(z:q_i,q_j)e_{q_i,q_j}(xz),$$
$$\mathcal{M}(x,z:q_i,q_j) := \mathcal{M}(z:q_i,q_j)e_{q_i,q_j}(xz).$$

By using Definition 1 and Corollary 7, it becomes

$$\sum_{n=0}^{\infty} \mathcal{B}_n(x:q_i,q_j)\frac{z^n}{[n]_{q_i,q_j}!}$$
$$=\left(\sum_{n=0}^{\infty} \mathcal{B}_n(q_i,q_j)\frac{z^n}{[n]_{q_i,q_j}!}\right)\left(\sum_{n=0}^{\infty} x^n\frac{z^n}{[n]_{q_i,q_j}!}\right).$$

From the rule of Cauchy product, we get

$$\sum_{n=0}^{\infty} \mathcal{B}_n(x:q_i,q_j)\frac{z^n}{[n]_{q_i,q_j}!}$$
$$=\sum_{n=0}^{\infty}\left(\sum_{k=0}^{\infty}\begin{bmatrix} n \\ k \end{bmatrix}_{q_i,q_j} \mathcal{B}_k(q_i,q_j)x^{n-k}\right)\frac{z^n}{[n]_{q_i,q_j}!}. \quad (2.2)$$

Comparing the coefficients of $\frac{z^n}{[n]_{q_i,q_j}!}$ in (2.2), we have

$$\mathcal{B}_n(x:q_i,q_j)=\sum_{k=0}^{\infty}\begin{bmatrix} n \\ k \end{bmatrix}_{q_i,q_j} \mathcal{B}_k(q_i,q_j)x^{n-k}. \quad (2.3)$$

From this, we can get similar identities for Euler-type and Genocchi-type polynomials. Therefore, we state the following theorem.

**Theorem 1.** *The following identities hold true:*

$$\mathcal{B}_n(x:q_i,q_j)=\sum_{k=0}^{\infty}\begin{bmatrix} n \\ k \end{bmatrix}_{q_i,q_j} \mathcal{B}_k(q_i,q_j)x^{n-k},$$

$$\mathcal{E}_n(x:q_i,q_j)=\sum_{k=0}^{\infty}\begin{bmatrix} n \\ k \end{bmatrix}_{q_i,q_j} \mathcal{E}_k(q_i,q_j)x^{n-k},$$

$$\mathcal{G}_n(x:q_i,q_j)=\sum_{k=0}^{\infty}\begin{bmatrix} n \\ k \end{bmatrix}_{q_i,q_j} \mathcal{G}_k(q_i,q_j)x^{n-k}.$$

Now we are in a position to investigate some properties of Bernoulli-type numbers and polynomials, Euler-type numbers and polynomials and Genocchi-type numbers and polynomials as follows.

From Definition 1 and by using Cauchy product, we get

$$\frac{z}{e_{q_i,q_j}(z)-1}e_{q_i,q_j}(z)=z+\frac{z}{e_{q_i,q_j}(z)-1}$$

$$\left(\sum_{n=0}^{\infty}\mathcal{B}_n(q_i,q_j)\frac{z^n}{[n]_{q_i,q_j}!}\right)\left(\sum_{n=0}^{\infty}\frac{z^n}{[n]_{q_i,q_j}!}\right)\sum_{n=0}^{\infty}\left(\sum_{k=0}^{\infty}\begin{bmatrix}n\\k\end{bmatrix}_{q_i,q_j}\mathcal{B}_n(q_i,q_j)\right)\frac{z^n}{[n]_{q_i,q_j}!}$$

$$=z+\sum_{n=0}^{\infty}\mathcal{B}_n(q_i,q_j)\frac{z^n}{[n]_{q_i,q_j}!} \qquad =z+\sum_{n=0}^{\infty}\mathcal{B}_n(q_i,q_j)\frac{z^n}{[n]_{q_i,q_j}!}$$

If we compute both of side and then compare coefficent of $\frac{z^n}{[n]_{q_i,q_j}!}$, then for $n > 1$, we acquire

$$\sum_{k=0}^{\infty}\frac{\mathcal{B}_k(q_i,q_j)}{[k]_{q_i,q_j}![n-k]_{q_i,q_j}!}-\frac{\mathcal{B}_n(q_i,q_j)}{[n]_{q_i,q_j}!}=0$$

$$\sum_{k=0}^{\infty}\begin{bmatrix}n\\k\end{bmatrix}_{q_i,q_j}\mathcal{B}_k(q_i,q_j)=\mathcal{B}_n(q_i,q_j).$$

(2.4)From this, we can get similar identities for Euler-type numbers and Genocchi-type numbers. The following theorem is an immediate consequence of Eq. (2.4).

**Theorem 2.** *(Recurrence Formula) For $n > 1$, we have*

$$\sum_{k=0}^{n-1}\begin{bmatrix} n \\ k \end{bmatrix}_{q_i,q_j} \mathcal{B}_k(q_i,q_j)=\begin{cases}1, & n=1\\ 0, & n>1\end{cases}$$

$$\sum_{k=0}^{\infty}\begin{bmatrix} n \\ k \end{bmatrix}_{q_i,q_j} \mathcal{E}_k(q_i,q_j)+\mathcal{E}_n(q_i,q_j)=\begin{cases}[2]_{q_i,q_j}, & n=0\\ 0, & n\geq 1\end{cases}$$

$$\sum_{k=0}^{\infty}\begin{bmatrix} n \\ k \end{bmatrix}_{q_i,q_j} \mathcal{G}_k(q_i,q_j)+\mathcal{G}_n(q_i,q_j)=\begin{cases}[2]_{q_i,q_j}, & n=1\\ 0, & n>1\end{cases}.$$

It is not diffucult to show the following equality:

$$\begin{bmatrix} n \\ k \end{bmatrix}_{q_i,q_j}\begin{bmatrix} n-k \\ m \end{bmatrix}_{q_i,q_j}=\begin{bmatrix} n \\ m+k \end{bmatrix}_{q_i,q_j}\begin{bmatrix} m+k \\ m \end{bmatrix}_{q_i,q_j}. \tag{2.5}$$

By (2.5), we get readily the following theorem.

**Theorem 3.** *For* $n\in\mathbb{N}$, *the followings hold true*

$$\mathcal{B}_n(x+y:q_i,q_j)=\sum_{k=0}^{n}\begin{bmatrix} n \\ k \end{bmatrix}_{q_i,q_j} \mathcal{B}_k(x:q_i,q_j)y^{n-k},$$

$$\mathcal{E}_n(x+y:q_i,q_j)=\sum_{k=0}^{n}\begin{bmatrix} n \\ k \end{bmatrix}_{q_i,q_j} \mathcal{E}_k(x:q_i,q_j)y^{n-k},$$

$$\mathcal{G}_n(x+y:q_i,q_j)=\sum_{k=0}^{n}\begin{bmatrix} n \\ k \end{bmatrix}_{q_i,q_j} \mathcal{G}_k(x:q_i,q_j)y^{n-k}.$$

*Proof.* If we change *x* by *x* + *y* in $\mathcal{S}_n(x,z:q_i,q_j)$, then we acquire

$$\sum_{n=0}^{\infty}\mathcal{B}_n(x+y:q_i,q_j)\frac{z^n}{[n]_{q_i,q_j}!}$$

$$=\frac{z}{e_{q_i,q_j}(z)-1}e_{q_i,q_j}((x+y)z)$$

$$=\sum_{n=0}^{\infty}\mathcal{B}_n(q_i,q_j)\frac{z^n}{[n]_{q_i,q_j}!}\sum_{n=0}^{\infty}\frac{z^n}{[n]_{q_i,q_j}!}$$

$$=\sum_{n=0}^{\infty}\left(\sum_{k=0}^{\infty}\begin{bmatrix} n \\ k \end{bmatrix}_{q_i,q_j}\mathcal{B}_n(x:q_i,q_j)y^{n-k}\right)\frac{z^n}{[n]_{q_i,q_j}!}.$$

By computing the coefficient $\frac{z^n}{[n]_{q_i,q_j}!}$ of both of side, then we have

$$\mathcal{B}_n(x+y:q_i,q_j)=\sum_{k=0}^{\infty}\begin{bmatrix} n \\ k \end{bmatrix}_{q_i,q_j}\mathcal{B}_n(x:q_i,q_j)y^{n-k}.$$

The others can be proved in a like manner.

Now we consider the special cases of Theorem 3 as Corollary 8 and Corollary 9.

**Corollary 8.** *Letting y = 1 in the Theorem* 3, *we then get*

$$\mathcal{B}_n(x+1:q_i,q_j)=\sum_{k=0}^{\infty}\begin{bmatrix} n \\ k \end{bmatrix}_{q_i,q_j}\mathcal{B}_k(x:q_i,q_j),$$

$$\mathcal{E}_n(x+1:q_i,q_j)=\sum_{k=0}^{\infty}\begin{bmatrix} n \\ k \end{bmatrix}_{q_i,q_j}\mathcal{E}_k(x:q_i,q_j),$$

$$\mathcal{G}_n(x+1:q_i,q_j)=\sum_{k=0}^{\infty}\begin{bmatrix} n \\ k \end{bmatrix}_{q_i,q_j}\mathcal{G}_k(x:q_i,q_j).$$

**Corollary 9.** *Letting x* = 0 *in the Theorem 3, we then get*

$$\mathcal{B}_n(x:q_i,q_j)=\sum_{k=0}^{\infty}\begin{bmatrix} n \\ k \end{bmatrix}_{q_i,q_j}\mathcal{B}_k(q_i,q_j)x^{n-k},$$

$$\mathcal{E}_n(x:q_i,q_j)=\sum_{k=0}^{\infty}\begin{bmatrix} n \\ k \end{bmatrix}_{q_i,q_j}\mathcal{E}_k(q_i,q_j)x^{n-k},$$

$$\mathcal{G}_n(x:q_i,q_j)=\sum_{k=0}^{\infty}\begin{bmatrix} n \\ k \end{bmatrix}_{q_i,q_j}\mathcal{G}_k(q_i,q_j)x^{n-k}.$$

**Theorem 4.** *The following expressions hold true for* $n \in \mathbb{N}$

$$\mathcal{B}_n\left(x+1 : q_i, q_j\right) - \mathcal{B}_n\left(x : q_i, q_j\right) = [n]_{q_i, q_j} x^{n-1}$$

$$\mathcal{E}_n\left(x+1 : q_i, q_j\right) + \mathcal{E}_n\left(x : q_i, q_j\right) + = [2]_{q_i, q_j} x^{n}$$

$$\mathcal{G}_n\left(x+1 : q_i, q_j\right) + \mathcal{G}_n\left(x : q_i, q_j\right) = [2]_{q_i, q_j} [n]_{q_i, q_j} x^{n-1}.$$

*Proof.* By using definitions of these polynomials and numbers, one can easily obtain these relations.

**Theorem 5.** *(Identity of Symmetry) The followings hold true for* $n \in \mathbb{N}$*:*

$$\mathcal{B}_n\left(1-x : q_i, q_j\right) = (-1)^n \mathcal{B}_n\left(x : q_i, q_j\right),$$

$$\mathcal{E}_n\left(1-x : q_i, q_j\right) = (-1)^n \mathcal{E}_n\left(x : q_i, q_j\right),$$

$$\mathcal{G}_n\left(1-x : q_i, q_j\right) = (-1)^{n+1} \mathcal{G}_n\left(x : q_i, q_j\right).$$

*Proof.* Setting $1-x$ instead of *x* in $\mathcal{S}_n\left(x, z : q_i, q_j\right)$, we then get

$$\frac{z}{e_{q_i, q_j}(z) - 1} e_{q_i, q_j}\left((1-x)z\right)$$

$$= \sum_{n=0}^{\infty} \mathcal{B}_n\left(1-x : q_i, q_j\right) \frac{z^n}{[n]_{q_i, q_j}!}$$

$$= \frac{(-z)}{e_{q_i, q_j}(-z) - 1} e_{q_i, q_j}\left(x(-z)\right)$$

$$= \sum_{n=0}^{\infty} (-1)^n \mathcal{B}_n\left(x : q_i, q_j\right) \frac{z^n}{[n]_{q_i, q_j}!}.$$

Compairing coefficients $\frac{z^n}{[n]_{q_i, q_j}!}$ both of side in above equality, we have desired the result. Similar to that of this proof, it can be proved for

Euler-type polynomials and Genocchi-type polynomials. So we completed this proof.

**Theorem 6.** *(Raabe's Formula) For* $n \in \mathbb{N}$, *the followings hold true*

$$\mathcal{B}_n\left(dx : q_i, q_j\right) = d^{n-1} \sum_{k=0}^{d-1} \mathcal{B}_n\left(x + \frac{k}{d} : q_i, q_j\right) \quad \left(d \in \mathbb{Z}^+\right)$$

$$\mathcal{E}_n\left(dx : q_i, q_j\right)$$
$$= d^n \sum_{k=0}^{d-1} (-1)^k \mathcal{E}_n\left(x + \frac{k}{d} : q_i, q_j\right) \quad (d \equiv 1 \pmod 2)$$

$$\mathcal{G}_n\left(dx : q_i, q_j\right)$$
$$= d^{n-1} \sum_{k=0}^{d-1} (-1)^k \mathcal{G}_n\left(x + \frac{k}{d} : q_i, q_j\right) \quad (d \equiv 1 \pmod 2).$$

*Proof.* By using Definition 1, then we have

$$\sum_{n=0}^{\infty}\left(d^{n-1} \sum_{k=0}^{d-1} \mathcal{B}_n\left(x + \frac{k}{d} : q_i, q_j\right)\right) \frac{z^n}{[n]_{q_i, q_j}!}$$
$$= \frac{1}{d} \sum_{k=0}^{d-1} \sum_{n=0}^{\infty} \mathcal{B}_n\left(x + \frac{k}{d} : q_i, q_j\right) \frac{(dz)^n}{[n]_{q_i, q_j}!}$$
$$= \frac{1}{d} \sum_{k=0}^{d-1} \frac{dz}{e_{q_i, q_j}(dz) - 1} e_{q_i, q_j}\left(\left(x + \frac{k}{d}\right) dz\right)$$
$$= \frac{z}{e_{q_i, q_j}(dz) - 1} e_{q_i, q_j}(xdz) \sum_{k=0}^{d-1} e_{q_i, q_j}(kz)$$
$$= \sum_{n=0}^{\infty} \mathcal{B}_n\left(dx : q_i, q_j\right) \frac{z^n}{[n]_{q_i, q_j}!}.$$

Similarly, we can prove this theorem for Euler-type numbers and Genocchi-type polynomials. So we omit them. Hence, we complete the proof of this theorem.

**Theorem 7.** *The three relations between Euler-type numbers and polynomials and Genocchi-type numbers and polynomials are given by*

$$\mathcal{E}_n(q_i,q_j)=\frac{\mathcal{G}_{n+1}(q_i,q_j)}{n+1},$$

$$\mathcal{E}_n(x:q_i,q_j)=\frac{\mathcal{G}_{n+1}(x:q_i,q_j)}{n+1},$$

$$\mathcal{E}_n(x:q_i,q_j)=\sum_{k=0}^{\infty}\begin{bmatrix} n \\ k \end{bmatrix}_{q_i,q_j}\frac{\mathcal{G}_{k+1}(q_i,q_j)}{k+1}x^{n-k}.$$

*Proof.* By Definition 1, we can easily obtain these relations. So we omit the proof.

Let us now apply the multiple $q$-derivative $D_{q_i,q_j}$, with respect to $x$, on the both sides of Definition 1,

$$D_{q_i,q_j}\left(\sum_{n=0}^{\infty}\mathcal{B}_n(x:q_i,q_j)\frac{z^n}{[n]_{q_i,q_j}!}\right)$$

$$=\sum_{n=0}^{\infty}D_{q_i,q_j}\mathcal{B}_n(x:q_i,q_j)\frac{z^n}{[n]_{q_i,q_j}!}$$

$$=\frac{z}{e_{q_i,q_j}(z)-1}D_{q_i,q_j}e_{q_i,q_j}(xz)$$

$$=\frac{z^2}{e_{q_i,q_j}(z)-1}e_{q_i,q_j}(xz)$$

$$=\sum_{n=0}^{\infty}\mathcal{B}_n(x:q_i,q_j)\frac{z^{n+1}}{[n]_{q_i,q_j}!}.$$

Matching the coefficients of $\frac{z^n}{[n]_{q_i,q_j}!}$ gives us

$$D_{q_i,q_j}\mathcal{B}_n(x:q_i,q_j)=[n]_{q_i,q_j}\mathcal{B}_{n-1}(x:q_i,q_j).$$

Thus we procure the following theorem.

**Theorem 8.** *The following identities hold true:*

$$D_{q_i,q_j}\mathcal{B}_n(x:q_i,q_j)=[n]_{q_i,q_j}\mathcal{B}_{n-1}(x:q_i,q_j),$$
$$D_{q_i,q_j}\mathcal{E}_n(x:q_i,q_j)=[n]_{q_i,q_j}\mathcal{E}_{n-1}(x:q_i,q_j)$$

*And*

$$D_{q_i,q_j}\mathcal{G}_n(x:q_i,q_j)=[n]_{q_i,q_j}\mathcal{G}_{n-1}(x:q_i,q_j).$$

Applying $k$-times the operator $D_{q_i,q_j}$ denoted by $D^k_{q_i,q_j}$ and the limit $t\to 0,$ respectively, to the Definition 1, we derive that

$$\mathcal{B}_n(x:q_i,q_j)=\lim_{t\to 0}D^k_{q_i,q_j}\frac{ze_{q_i,q_j}(xz)}{e_{q_i,q_j}(z)-1}.$$

So we conclude the following theorem.

**Theorem 9.** *For* $k\geq 0$ *and* $n\geq 0$*, we have*

$$\mathcal{B}_n(x:q_i,q_j)=\lim_{t\to 0}D^k_{q_i,q_j}\frac{ze_{q_i,q_j}(xz)}{e_{q_i,q_j}(z)-1},$$
$$\mathcal{E}_n(x:q_i,q_j)=\lim_{t\to 0}D^k_{q_i,q_j}\frac{[2]_{q_i,q_j}}{e_{q_i,q_j}(z)+1}e_{q_i,q_j}(xz)$$

*And*

$$\mathcal{G}_n(x:q_i,q_j)=\lim_{t\to 0}D^k_{q_i,q_j}\frac{[2]_{q_i,q_j}z}{e_{q_i,q_j}(z)+1}e_{q_i,q_j}(xz).$$

**Definition 2.** *Let* $0<a<b$. *The definite multiple q-integral has the following representation:*

$$\int_a^b f\left(\frac{x}{q_i}\right)d_{\frac{q_j}{q_i}}x=(q_i-q_j)b\sum_{k=0}^{\infty}\frac{q_j^k}{q_i^{k+1}}f\left(\frac{q_j^k}{q_i^{k+1}}b\right)$$

*And*

$$\int_a^b f\left(\frac{x}{q_i}\right) d_{\frac{q_j}{q_i}} x = \int_0^b f\left(\frac{x}{q_i}\right) d_{\frac{q_j}{q_i}} x - \int_0^a f\left(\frac{x}{q_i}\right) d_{\frac{q_j}{q_i}} x.$$

**Theorem 10.** *The following holds true:*

$$\int_0^b f\left(\frac{x}{q_i}\right) D_{q_i,q_j} g\left(\frac{x}{q_i}\right) d_{\frac{q_j}{q_i}} x$$

$$= \sum_{n=0}^{\infty} f\left(\frac{q_j^k}{q_i^{k+1}} b\right)\left(g\left(\frac{q_j^k b}{q_i^k}\right) - g\frac{q_j^{k+1} b}{q_i^{k+1}}\right)$$

*Proof.* From Definition 2, we write that

$$\int_0^b f\left(\frac{x}{q_i}\right) D_{q_i,q_j} g\left(\frac{x}{q_i}\right) d_{\frac{q_j}{q_i}} x$$

$$= \left(q_i - q_j\right) b \sum_{k=0}^{\infty} \frac{q_j^k}{q_i^{k+1}} f\left(\frac{q_j^k}{q_i^{k+1}} b\right) D_{q_i,q_j} g\left(\frac{q_j^k}{q_i^{k+1}} b\right), \tag{2.6}$$

*where* $D_{q_i,q_j} g\left(\frac{q_j^k}{q_i^{k+1}} b\right)$ *equals to*

$$\frac{g\left(\frac{q_j^k}{q_i^k} b\right) - g\left(\frac{q_j^{k+1}}{q_i^{k+1}} b\right)}{\left(q_i - q_j\right) b}. \tag{2.7}$$

Combining the Eq. (2.6) with the Eq. (2.7) gives us the proof of the theorem.

**Theorem 11.** $\left|\frac{q_j}{q_i}\right| < 1$ *and* $i, j \in \{1, 2, \ldots, N\}$. *Then we have*

$$\int_0^1 \mathcal{B}_n\left(\frac{x}{q_i} : q_i, q_j\right) d_{\frac{q_j}{q_i}} x = \frac{\mathcal{B}_{n+1}(1 : q_i, q_j) - \mathcal{B}_{n+1}(q_i, q_j)}{[n+1]_{q_i, q_j}}$$

$$\int_0^1 \mathcal{E}_n\left(\frac{x}{q_i} : q_i, q_j\right) d_{\frac{q_j}{q_i}} x = \frac{\mathcal{E}_{n+1}(1 : q_i, q_j) - \mathcal{E}_{n+1}(q_i, q_j)}{[n+1]_{q_i, q_j}}$$

$$\int_0^1 \mathcal{G}_n\left(\frac{x}{q_i} : q_i, q_j\right) d_{\frac{q_j}{q_i}} x = \frac{\mathcal{G}_{n+1}(1 : q_i, q_j) - \mathcal{G}_{n+1}(q_i, q_j)}{[n+1]_{q_i, q_j}}.$$

*Proof.* By using Theorem 1, Definition 2 and for $\left|\frac{q_j}{q_i}\right| < 1$, we have

$$\int_0^1 \mathcal{B}_n\left(\frac{x}{q_i} : q_i, q_j\right) d_{\frac{q_j}{q_i}} x$$

$$= \sum_{l=0}^{n} \begin{bmatrix} n \\ l \end{bmatrix}_{q_i, q_j} \mathcal{B}_{n-l}(q_i, q_j) \left( (q_i - q_j) \sum_{k=0}^{\infty} \frac{q_j^k}{q_i^{k+1}} \left( \frac{q_j^k}{q_i^{k+1}} \right)^l \right)$$

$$= \sum_{l=0}^{n} \begin{bmatrix} n \\ l \end{bmatrix}_{q_i, q_j} \mathcal{B}_{n-l}(q_i, q_j) \frac{1}{[l+1]_{q_i, q_j}}$$

$$= \frac{\mathcal{B}_{n+1}(1 : q_i, q_j) - \mathcal{B}_{n+1}(q_i, q_j)}{[n+1]_{q_i, q_j}}.$$

Similarly, the identities of Euler-type polynomials and Genocchi-type polynomials can be shown. Therefore, we complete the proof of theorem.

# 3 FURTHER REMARKS

Here we list a few values of Bernoulli-type, Euler-type and Genocchi-type numbers as follows:

**Bernoulli-type number:**

**Table 1.**

| |
|---|
| $\mathcal{B}_0(q_i,q_j)=1$ |
| $\mathcal{B}_1(q_i,q_j)=-\frac{1}{[2]_{q_i,q_j}}$ |
| $\mathcal{B}_2(q_i,q_j)=-\frac{1}{[3]_{q_i,q_j}}+\frac{1}{[2]_{q_i,q_j}}$ |
| $\mathcal{B}_3(q_i,q_j)=-\frac{1}{[4]_{q_i,q_j}}+\frac{2}{[2]_{q_i,q_j}}-\frac{[3]_{q_i,q_j}}{[2]_{q_i,q_j}^2}$ |
| $\mathcal{B}_4(q_i,q_j)=-\frac{1}{[5]_{q_i,q_j}}+\frac{1}{[2]_{q_i,q_j}}+\frac{[4]_{q_i,q_j}}{[2]_{q_i,q_j}}\left(\frac{1}{[3]_{q_i,q_j}}-\frac{1}{[2]_{q_i,q_j}}\right)$ |
| $+\frac{[4]_{q_i,q_j}}{[2]_{q_i,q_j}}\left(\frac{1}{[4]_{q_i,q_j}}-\frac{2}{[2]_{q_i,q_j}}+\frac{[3]_{q_i,q_j}}{[2]_{q_i,q_j}^2}\right)$ |

As a special case of Table 1, we get

| |
|---|
| $\mathfrak{B}_0(q)=1$ |
| $\mathfrak{B}_1(q)=-\frac{1}{[2]_q}$ |
| $\mathfrak{B}_2(q)=-\frac{1}{[3]_q}+\frac{1}{[2]_q}$ |
| $\mathfrak{B}_3(q)=-\frac{1}{[4]_q}+\frac{2}{[2]_q}-\frac{[3]_q}{[2]_q^2}$ |
| $\mathfrak{B}_4(q)=-\frac{1}{[5]_q}+\frac{1}{[2]_q}+\frac{[4]_q}{[2]_q}\left(\frac{1}{[3]_q}-\frac{1}{[2]_q}\right)$ |
| $+\frac{[4]_q}{[2]_q}\left(\frac{1}{[4]_q}-\frac{2}{[2]_q}+\frac{[3]_q}{[2]_q^2}\right)$ |

As a special case of Table 1, we get

| |
|---|
| $B_0 = 1$ |
| $B_1 = -\frac{1}{2}$ |
| $B_2 = \frac{1}{6}$ |
| $B_3 = 0$ |
| $B_4 = -\frac{1}{30}$ |

Moreover the first few Bernoulli-type numbers can be shown $N \times N$ matrix with multiple $q$-numbers elements in the following form

$$\left(\mathcal{B}_0\left(q_i, q_j\right)\right) = \begin{pmatrix} 1 & 1 & \dots & 1 \\ 1 & 1 & \dots & 1 \\ \dots & \dots & \dots & \dots \\ 1 & 1 & \dots & 1 \end{pmatrix}$$

$$\left(\mathcal{B}_1\left(q_i, q_j\right)\right) = \begin{pmatrix} -\frac{1}{2q_1} & -\frac{1}{[2]_{q_1,q_2}} & \dots & -\frac{1}{[2]_{q_1,q_N}} \\ -\frac{1}{[2]_{q_2,q_1}} & -\frac{1}{2q_2} & \dots & -\frac{1}{[2]_{q_2,q_N}} \\ \dots & \dots & \dots & \dots \\ -\frac{1}{[2]_{q_N,q_1}} & -\frac{1}{[2]_{q_N,q_1}} & \dots & -\frac{1}{[2]q_N} \end{pmatrix}$$

$$\left(\mathcal{B}_2(q_i,q_j)\right)=\begin{pmatrix} \left(-\frac{1}{3q_1^2}+\frac{1}{2q_1}\right) & \left(-\frac{1}{[3]_{q_1,q_2}}+\frac{1}{[2]_{q_1,q_2}}\right) & \cdots & \left(-\frac{1}{[3]_{q_1,q_N}}+\frac{1}{[2]_{q_1,q_N}}\right) \\ \left(-\frac{1}{[3]_{q_2,q_1}}+\frac{1}{[2]_{q_2,q_1}}\right) & \left(-\frac{1}{3q_2^2}+\frac{1}{2q_2}\right) & \cdots & \left(-\frac{1}{[3]_{q_2,q_N}}+\frac{1}{[2]_{q_2,q_N}}\right) \\ \cdots & \cdots & \cdots & \cdots \\ \left(-\frac{1}{[3]_{q_N,q_1}}+\frac{1}{[2]_{q_N,q_1}}\right) & \left(-\frac{1}{[3]_{q_N,q_2}}+\frac{1}{[2]_{q_N,q_2}}\right) & \cdots & \left(-\frac{1}{3q_N^2}+\frac{1}{2q_N}\right) \end{pmatrix}$$

From Definition 1 and the Table 1, we easily acquire the first few Bernoulli-type polynomials

**Bernoulli-type polynomials**

$$\mathcal{B}_0(x:q_i,q_j)=1$$

$$\mathcal{B}_1(x:q_i,q_j)=x-\frac{1}{[2]_{q_i,q_j}}$$

$$\mathcal{B}_2(x:q_i,q_j)=x^2-x-\frac{1}{[3]_{q_i,q_j}}+\frac{1}{[2]_{q_i,q_j}}$$

**Usual Bernoulli polynomials**

$$B_0=1$$

$$B_1=x-\frac{1}{2}$$

$$B_2=x^2-x-\frac{1}{6}$$

Moreover, the first few Bernoulli-type polynomials can be shown $N\times N$ matrix with $q$-numbers elements in the following form

$$\left(\mathcal{B}_0\left(x:q_i,q_j\right)\right)=\begin{pmatrix} 1 & 1 & \cdots & 1 \\ 1 & 1 & \cdots & 1 \\ \cdots & \cdots & \cdots & \cdots \\ 1 & 1 & \cdots & 1 \end{pmatrix}=\left(\mathcal{B}_0\left(q_i,q_j\right)\right)$$

$$\left(\mathcal{B}_1\left(x:q_i,q_j\right)\right)$$
$$=\begin{pmatrix} x-\frac{1}{2q_1} & x-\frac{1}{[2]_{q_1,q_2}} & \cdots & x-\frac{1}{[2]_{q_1,q_N}} \\ x-\frac{1}{[2]_{q_2,q_1}} & x-\frac{1}{2q_2} & \cdots & x-\frac{1}{[2]_{q_2,q_N}} \\ \cdots & \cdots & \cdots & \cdots \\ x-\frac{1}{[2]_{q_N,q_1}} & x-\frac{1}{[2]_{q_N,q_1}} & \cdots & x-\frac{1}{[2]q_N} \end{pmatrix}$$

$$\left(\mathcal{B}_2\left(x:q_i,q_j\right)\right)=\begin{pmatrix} x^2-x-\frac{1}{3q_1^2}+\frac{1}{2q_1} & x^2-x-\frac{1}{[3]_{q_1,q_2}}+\frac{1}{[2]_{q_1,q_2}} & \cdots & x^2-x-\frac{1}{[3]_{q_1,q_N}}+\frac{1}{[2]_{q_1,q_N}} \\ x^2-x-\frac{1}{[3]_{q_2,q_1}}+\frac{1}{[2]_{q_2,q_1}} & x^2-x-\frac{1}{3q_2^2}+\frac{1}{2q_2} & \cdots & x^2-x-\frac{1}{[3]_{q_2,q_N}}+\frac{1}{[2]_{q_2,q_N}} \\ \cdots & \cdots & \cdots & \cdots \\ x^2-x-\frac{1}{[3]_{q_N,q_1}}+\frac{1}{[2]_{q_N,q_1}} & x^2-x-\frac{1}{[3]_{q_N,q_2}}+\frac{1}{[2]_{q_N,q_2}} & \cdots & x^2-x-\frac{1}{3q_N^2}+\frac{1}{2q_N} \end{pmatrix}$$

**Euler-type Numbers and Polynomials**:

We begin to compute the first few value of $\mathcal{E}_n\left(q_i,q_j\right)$ as follows:

**Table 2.**

| |
|---|
| $\mathcal{E}_0(q_i,q_j)=\frac{[2]_{q_i,q_j}}{2}$ |
| $\mathcal{E}_1(q_i,q_j)=-\frac{[2]_{q_i,q_j}}{4}$ |
| $\mathcal{E}_2(q_i,q_j)=-\frac{[2]_{q_i,q_j}}{4}+\frac{\left([2]_{q_i,q_j}\right)^2}{8}$ |
| $\mathcal{E}_3(q_i,q_j)=-\frac{[2]_{q_i,q_j}}{4}+\frac{[3]_{q_i,q_j}[2]_{q_i,q_j}}{4}\left(1-\frac{[2]_{q_i,q_j}}{4}\right)$ |
| $\mathcal{E}_4(q_i,q_j)=-\frac{[2]_{q_i,q_j}}{4}+\frac{[4]_{q_i,q_j}[2]_{q_i,q_j}}{4}+\frac{[4]_{q_i,q_j}[3]_{q_i,q_j}[2]_{q_i,q_j}}{8}\left(\frac{1}{[2]_{q_i,q_j}}-\frac{3}{2}+\frac{[2]_{q_i,q_j}}{4}\right)$ |
| $\mathcal{E}_5(q_i,q_j)=-\frac{[2]_{q_i,q_j}}{4}+\frac{[5]_{q_i,q_j}[2]_{q_i,q_j}}{4}+\frac{[5]_{q_i,q_j}[4]_{q_i,q_j}[2]_{q_i,q_j}}{4}\left(-\frac{3}{4}-\frac{1}{[2]_{q_i,q_j}}\right)$ |
| $+\frac{[5]_{q_i,q_j}[4]_{q_i,q_j}[3]_{q_i,q_j}[2]_{q_i,q_j}}{16}\left(\frac{3}{[2]_{q_i,q_j}}-2+\frac{[2]_{q_i,q_j}}{4}\right)$ |

As a special case of Table 2, we have

| |
|---|
| $E_0(q)=\frac{[2]_q}{2}$ |
| $E_1(q)=-\frac{[2]_q}{4}$ |
| $E_2(q)=-\frac{[2]_q}{4}+\frac{\left([2]_q\right)^2}{8}$ |
| $E_3(q)=-\frac{[2]_q}{4}+\frac{[3]_q[2]_q}{4}\left(1-\frac{[2]_q}{4}\right)$ |
| $E_4(q)=-\frac{[2]_q}{4}+\frac{[4]_q[2]_q}{4}+\frac{[4]_q[3]_q[2]_q}{8}\left(\frac{1}{[2]_q}-\frac{3}{2}+\frac{[2]_q}{4}\right)$ |
| $E_5(q)=-\frac{[2]_q}{4}+\frac{[5]_q[2]_q}{4}+\frac{[5]_q[4]_q[2]_q}{4}\left(-\frac{3}{4}-\frac{1}{[2]_q}\right)$ |
| $+\frac{[5]_q[4]_q[3]_q[2]_q}{16}\left(\frac{3}{[2]_q}-2+\frac{[2]_q}{4}\right)$ |

As a special case of Table 2, we have

| |
|---|
| $E_0(q)=1$ |
| $E_1(q)=-\frac{1}{2}$ |
| $E_2(q)=0$ |
| $E_3(q)=\frac{1}{4}$ |
| $E_4(q)=0$ |
| $E_5(q)=-\frac{1}{2}$ |

Moreover the first few Euler-type numbers can be shown $N\times N$ matrix with $q$-numbers elements in the following form

$$\left(\mathcal{E}_0(q_i,q_j)\right)=\begin{pmatrix} q_1 & \frac{[2]_{q_1,q_2}}{2} & \cdots & \frac{[2]_{q_1,q_N}}{2} \\ \frac{[2]_{q_2,q_1}}{2} & q_2 & \cdots & \frac{[2]_{q_2,q_N}}{2} \\ \cdots & \cdots & \cdots & \cdots \\ \frac{[2]_{q_N,q_1}}{2} & \frac{[2]_{q_N,q_2}}{2} & \cdots & q_N \end{pmatrix}$$

$$\left(\mathcal{E}_1(q_i,q_j)\right)=\begin{pmatrix} -\frac{q_1}{2} & -\frac{[2]_{q_1,q_2}}{4} & \cdots & -\frac{[2]_{q_1,q_N}}{4} \\ -\frac{[2]_{q_2,q_1}}{4} & -\frac{q_2}{2} & \cdots & -\frac{[2]_{q_2,q_N}}{4} \\ \cdots & \cdots & \cdots & \cdots \\ -\frac{[2]_{q_N,q_1}}{4} & -\frac{[2]_{q_N,q_2}}{4} & \cdots & -\frac{q_N}{2} \end{pmatrix}$$

$$\left(\mathcal{E}_2\left(x:q_i,q_j\right)\right)$$
$$=\begin{pmatrix} -\frac{q_1}{2}+\frac{q_1^2}{2} & \left(-\frac{[2]_{q_1,q_2}}{4}+\frac{\left([2]_{q_1,q_2}\right)^2}{8}\right) & \cdots & \left(-\frac{[2]_{q_1,q_N}}{4}+\frac{\left([2]_{q_1,q_N}\right)^2}{8}\right) \\ \left(-\frac{[2]_{q_2,q_1}}{4}+\frac{\left([2]_{q_2,q_1}\right)^2}{8}\right) & -\frac{q_2}{2}+\frac{q_2^2}{2} & \cdots & \left(-\frac{[2]_{q_2,q_N}}{4}+\frac{\left([2]_{q_2,q_N}\right)^2}{8}\right) \\ \cdots & \cdots & \cdots & \cdots \\ \left(-\frac{[2]_{q_N,q_1}}{4}+\frac{\left([2]_{q_N,q_1}\right)^2}{8}\right) & \left(-\frac{[2]_{q_N,q_2}}{4}+\frac{\left([2]_{q_N,q_2}\right)^2}{8}\right) & \cdots & -\frac{q_N}{2}+\frac{q_N^2}{2} \end{pmatrix}$$

From Definition 1 and the Table 2, we easily acquire the first few Euler-type polynomials

### Euler-type polynomials

$$\mathcal{E}_0\left(x:q_i,q_j\right)=\frac{[2]_{q_i,q_j}}{2}$$
$$\mathcal{E}_1\left(x:q_i,q_j\right)=\frac{[2]_{q_i,q_j}}{2}x-\frac{[2]_{q_i,q_j}}{4}$$
$$\mathcal{E}_2\left(x:q_i,q_j\right)=\frac{[2]_{q_i,q_j}}{2}x^2-\frac{\left([2]_{q_i,q_j}\right)^2}{4}x$$
$$-\frac{[2]_{q_i,q_j}}{4}+\frac{\left([2]_{q_i,q_j}\right)^2}{8}$$

### Usual Euler polynomials

$$E_0=1$$
$$E_1=x-\frac{1}{2}$$
$$E_2=x^2-x$$

Moreover the first few Euler-type polynomials can be shown $N \times N$ matrix with multiple $q$-numbers elements in the following form

$$
\left(\mathcal{E}_0\left(q_i, q_j\right)\right) = \begin{pmatrix} q_1 & \frac{[2]_{q_1,q_2}}{2} & \cdots & \frac{[2]_{q_1,q_N}}{2} \\ \frac{[2]_{q_2,q_1}}{2} & q_2 & \cdots & \frac{[2]_{q_2,q_N}}{2} \\ \cdots & \cdots & \cdots & \cdots \\ \frac{[2]_{q_N,q_1}}{2} & \frac{[2]_{q_N,q_2}}{2} & \cdots & q_N \end{pmatrix}
$$

$$
\left(\mathcal{E}_1\left(x : q_i, q_j\right)\right) = \begin{pmatrix} q_1 x - \frac{q_1}{2} & \frac{[2]_{q_1,q_2}}{2} x - \frac{[2]_{q_1,q_2}}{4} & \cdots & \frac{[2]_{q_1,q_2}}{2} x - \frac{[2]_{q_1,q_N}}{4} \\ \frac{[2]_{q_1,q_2}}{2} x - \frac{[2]_{q_2,q_1}}{4} & q_2 x - \frac{q_2}{2} & \cdots & \frac{[2]_{q_1,q_2}}{2} x - \frac{[2]_{q_2,q_N}}{4} \\ \cdots & \cdots & \cdots & \cdots \\ \frac{[2]_{q_1,q_2}}{2} x - \frac{[2]_{q_N,q_1}}{4} & \frac{[2]_{q_1,q_2}}{2} x - \frac{[2]_{q_N,q_2}}{4} & \cdots & q_N x - \frac{q_N}{2} \end{pmatrix}
$$

$$
\left(\mathcal{E}_2\left(x : q_i, q_j\right)\right) = \begin{pmatrix} q_1 x^2 - q_1^2 x - \frac{q_1}{2} + \frac{q_1^2}{2} & \cdots & \frac{[2]_{q_1,q_N}}{2} x^2 - \frac{\left([2]_{q_1,q_N}\right)^2}{4} x - \frac{[2]_{q_1,q_N}}{4} + \frac{\left([2]_{q_1,q_N}\right)^2}{8} \\ \frac{[2]_{q_1,q_2}}{2} x^2 - \frac{\left([2]_{q_2,q_1}\right)^2}{4} x - \frac{[2]_{q_1,q_2}}{4} + \frac{\left([2]_{q_1,q_2}\right)^2}{8} & \cdots & \frac{[2]_{q_2,q_N}}{2} x^2 - \frac{\left([2]_{q_2,q_N}\right)^2}{4} x - \frac{[2]_{q_1,q_N}}{4} + \frac{\left([2]_{q_1,q_N}\right)^2}{8} \\ \cdots & \cdots & \cdots \\ \frac{[2]_{q_N,q_1}}{2} x^2 - \frac{\left([2]_{q_N,q_1}\right)^2}{4} x - \frac{[2]_{q_N,q_1}}{4} + \frac{\left([2]_{q_N,q_1}\right)^2}{8} & \cdots & q_N x^2 - q_N^2 x - \frac{q_N}{2} + \frac{q_N^2}{2} \end{pmatrix}
$$

## Genocchi-type Numbers and Polynomials:

We begin to compute the first few value of $\mathcal{G}_1\left(q_i, q_j\right)$ as follows:

**Table 3.**

| |
|---|
| $\mathcal{G}_0(q_i,q_j)=0$ |
| $\mathcal{G}_1(q_i,q_j)=\frac{[2]_{q_i,q_j}}{2}$ |
| $\mathcal{G}_2(q_i,q_j)=-\frac{\left([2]_{q_i,q_j}\right)^2}{4}$ |
| $\mathcal{G}_3(q_i,q_j)=\frac{[3]_{q_i,q_j}}{4}\left(-[2]_{q_i,q_j}+\frac{\left([2]_{q_i,q_j}\right)^2}{2}\right)$ |
| $\mathcal{G}_4(q_i,q_j)=-\frac{[4]_{q_i,q_j}[2]_{q_i,q_j}}{4}+\frac{[4]_{q_i,q_j}[3]_{q_i,q_j}[2]_{q_i,q_j}}{8}-\frac{[4]_{q_i,q_j}[3]_{q_i,q_j}}{8}\left(-[2]_{q_i,q_j}+\frac{\left([2]_{q_i,q_j}\right)^2}{2}\right)$ |
| $\mathcal{G}_5(q_i,q_j)=-\frac{[5]_{q_i,q_j}[2]_{q_i,q_j}}{4}+\frac{[5]_{q_i,q_j}[4]_{q_i,q_j}[2]_{q_i,q_j}}{8}--\frac{[5]_{q_i,q_j}[4]_{q_i,q_j}[3]_{q_i,q_j}}{8}\left(-1+\frac{[2]_{q_i,q_j}}{2}\right)$ |
| $-\frac{[5]_{q_i,q_j}}{8}\left(-[4]_{q_i,q_j}[2]_{q_i,q_j}+\frac{[4]_{q_i,q_j}[3]_{q_i,q_j}[2]_{q_i,q_j}}{2}-\frac{[4]_{q_i,q_j}[3]_{q_i,q_j}}{2}\left(-[2]_{q_i,q_j}+\frac{\left([2]_{q_i,q_j}\right)^2}{2}\right)\right)$ |

As a special case of Table 3, we have

| |
|---|
| $\mathcal{E}_0(q)=0$ |
| $\mathcal{E}_1(q)=\frac{[2]_q}{2}$ |
| $\mathcal{E}_2(q)=\frac{\left([2]_q\right)^2}{4}$ |
| $\mathcal{E}_3(q)=\frac{[3]_q}{4}\left(-[2]_q+\frac{\left([2]_q\right)^2}{2}\right)$ |
| $\mathcal{E}_4(q)=-\frac{[4]_q[2]_q}{4}+\frac{[4]_q[3]_q[2]_q}{8}-\frac{[4]_q[3]_q}{8}\left(-[2]_q+\frac{\left([2]_q\right)^2}{2}\right)$ |
| $\mathcal{E}_5(q)=-\frac{[5]_q[2]_q}{4}+\frac{[5]_q[4]_q[2]_q}{8}--\frac{[5]_q[4]_q[3]_q}{8}\left(-1+\frac{[2]_q}{2}\right)$ |
| $-\frac{[5]_q}{8}\left(-[4]_q[2]_q+\frac{[4]_q[3]_q[2]_q}{2}-\frac{[4]_q[3]_q}{2}\left(-[2]_q+\frac{\left([2]_q\right)^2}{2}\right)\right)$ |

As a special case of Table 3, we have

| |
|---|
| $G_0 = 0$ |
| $G_1 = 1$ |
| $G_2 = -1$ |
| $G_3 = 0$ |
| $G_4 = 1$ |
| $G_5 = 0$ |

Moreover the first few Genocchi-type numbers can be shown $N \times N$ matrix with multiple $q$-numbers elements in the following form

$$\left(\mathcal{G}_0\left(q_i,q_j\right)\right)=\begin{pmatrix} 0 & 0 & \dots & 0 \\ 0 & 0 & \dots & 0 \\ \dots & \dots & \dots & \dots \\ 0 & 0 & \dots & 0 \end{pmatrix}$$

$$\left(\mathcal{G}_1\left(q_i,q_j\right)\right)=\begin{pmatrix} q_1 & \frac{[2]_{q_1,q_2}}{2} & \dots & \frac{[2]_{q_1,q_N}}{2} \\ \frac{[2]_{q_2,q_1}}{2} & q_2 & \dots & \frac{[2]_{q_2,q_N}}{2} \\ \dots & \dots & \dots & \dots \\ \frac{[2]_{q_N,q_1}}{2} & \frac{[2]_{q_N,q_2}}{2} & \dots & q_N \end{pmatrix}$$

$$\left(\mathcal{G}_2\left(q_i,q_j\right)\right)$$

$$=\begin{pmatrix} -\frac{q_1^2}{2} & -\frac{\left([2]_{q_1,q_2}\right)^2}{4} & \dots & -\frac{\left([2]_{q_1,q_N}\right)^2}{4} \\ -\frac{\left([2]_{q_2,q_1}\right)^2}{4} & -\frac{q_2^2}{2} & \dots & -\frac{\left([2]_{q_2,q_N}\right)^2}{4} \\ \dots & \dots & \dots & \dots \\ -\frac{\left([2]_{q_N,q_1}\right)^2}{4} & -\frac{\left([2]_{q_N,q_2}\right)^2}{4} & \dots & -\frac{q_N^2}{2} \end{pmatrix}$$

From Definition 1 and the Table 3, we easily acquire the first few Genocchi-type polynomials

**Genocchi-type polynomials**

$$\mathcal{G}_0\left(x:q_i,q_j\right)=0$$

$$\mathcal{G}_1\left(x:q_i,q_j\right)=\frac{[2]_{q_i,q_j}}{2}$$

$$\mathcal{G}_2\left(x:q_i,q_j\right)=\frac{\left([2]_{q_i,q_j}\right)^2}{2}x-\frac{\left([2]_{q_i,q_j}\right)^2}{4}$$

**Usual Genocchi polynomials**

$$G_0=0$$

$$G_1=1$$

$$G_2=2x-1$$

Moreover the first few Genocchi-type polynomials can be shown $N\times N$ matrix with multiple $q$-numbers elements in the following form

$$\left(\mathcal{G}_0\left(q_i,q_j\right)\right)=\begin{pmatrix}0 & 0 & \dots & 0\\ 0 & 0 & \dots & 0\\ \dots & \dots & \dots & \dots\\ 0 & 0 & \dots & 0\end{pmatrix}$$

$$(\mathcal{G}_1(q_i,q_j)) = \begin{pmatrix} q_1 & \frac{[2]_{q_1,q_2}}{2} & \cdots & \frac{[2]_{q_1,q_N}}{2} \\ \frac{[2]_{q_2,q_1}}{2} & q_2 & \cdots & \frac{[2]_{q_2,q_N}}{2} \\ \cdots & \cdots & \cdots & \cdots \\ \frac{[2]_{q_N,q_1}}{2} & \frac{[2]_{q_N,q_2}}{2} & \cdots & q_N \end{pmatrix}$$

$$(\mathcal{G}_2(x:q_i,q_j)) = \begin{pmatrix} 2q_1^2x - \frac{q_1^2}{2} & \frac{([2]_{q_1,q_2})^2}{2} - \frac{([2]_{q_1,q_2})^2}{4} & \cdots & \frac{([2]_{q_1,q_N})^2}{2} - \frac{([2]_{q_1,q_N})^2}{4} \\ \frac{([2]_{q_2,q_1})^2}{2} - \frac{([2]_{q_2,q_1})^2}{4} & 2q_2^2x - \frac{q_2^2}{2} & \cdots & \frac{([2]_{q_2,q_N})^2}{2} - \frac{([2]_{q_2,q_N})^2}{4} \\ \cdots & \cdots & \cdots & \cdots \\ \frac{([2]_{q_N,q_1})^2}{2} - \frac{([2]_{q_N,q_1})^2}{4} & \frac{([2]_{q_N,q_2})^2}{2} - \frac{([2]_{q_N,q_2})^2}{4} & \cdots & 2q_N^2x - \frac{q_N^2}{2} \end{pmatrix}$$

# REFERENCES


1. G. M. Philips, Bernstein polynomials based on the q-integers, Annals of Numerical Mathematics, vol. 4,no. 1-4, pp. 511-518, 1997.

2. C. S. Ryoo, A note on q-Bernoulli numbers and polynomials, Appl. Math. Lett. 20 (2007) 524-531.

3. V. Kac and P. Cheung, Quantum Calculus, New York: Springer, 2002.

4. G. S. Cheon, A note on the Bernoulli and Euler polynomials, Appl. Math. Lett. 16 (2003) 365-368.

5. Q. M. Luo, Some results for the q-Bernoulli and q- Euler polynomials, J. Math. Anal. Appl. 363 (2010) 7-18.

6. H. M. Srivastava, Some generalizations and basic (or q-) extensions of the Bernoulli, Euler and Genocchi polynomials, Appl. Math. Inform. Sci. 5 (2011), 390-444.

7. H. M. Srivastava, Some formulas for the Bernoulli and Euler polynomials at rational arguments, Math. Proc. Cambridge Philos. Soc. 129 (2000), 77-84.

8. S. Nalci and O. K. Pashaev, q-Bernoulli Numbers and Zeros of q-Sine Function, arXiv: 1202,2265v12012.

9. S. Nalci and O.K. Pashaev, Exactly Solvable q-extended Nonlinear Classical and Quantum Models, Lambert Academic Publishing, 2014.

10. M. Acikgoz, S. Araci, U.Duran, Research on Some New Theorems and Some Elementary FunctionsUnder Multiple Quantum Calculus, preprint.

11. U. Duran, M. Acikgoz, A. Esi, S. Araci, Some new symmetric identities involving q-Genocchi polynomialsunder S4, Journal of Mathematical Analysis, Vol 6, Issue 4 (2015).

12. U. Duran, M. Acikgoz, S. Araci, Symmetric identities involving weighted q-Genocchi polynomials underS4, Proceedings of the Jangjeon Mathematical Society,18 (2015), No. 4, pp 455-465.

13. S. Araci, E. Ağyüz and M. Acikgoz, On a q-analog of some numbers and polynomials, J. Ineq. Appl. (2015) 2015:19.

14. S. Araci, M. Acikgoz, A. Bagdasaryan, E. Sen, The Legendre Polynomials Associated with Bernoulli,Euler, Hermite and Bernstein Polynomials, Turkish J. Anal. Number Theory, Vol. 1, No. 1, (2013) 1-3.

15. S. Araci, M. Acikgoz, E. Sen, Some New Formulae for Genocchi Numbers and Polynomials Involving Bernoulli and Euler Polynomials, Int. J. Math. Math. Sci., vol. 2014(2014), Art. ID: 760613, 7pp.

16. S. Araci, U. Duran, M. Acikgoz, Symmetric Identities Involving q-Frobenius-Euler Polynomials under Sym (5), Turkish J. Anal. Number Theory, Vol. 3, No. 3, (2015) pp 90-93.

17. T. Kim, q-Volkenborn integration, Russ. J. Math. Phys. Vol. 9, no.3, pp. 288-299, 2002.

18. T. Kim, q-Generalized Euler numbers and polynomials, Russ. J. Math. Phys., Vol. 13, No. 3 (2006), pp. 293.298.

19. T. Kim, Symmetry p-adic invariant integral on  for Bernoulli and Euler polynomials, J. Differ. Equ. Appl. 14 (2008), 1267-1277.

20. B. A. Kupershmidt, Reflection Symmetries of q-Bernoulli Polynomials, J. Nonlinear Math. Phys. 12 (Suppl. 1), 412-422 (2005).

21. N. I. Mahmudov, On a class of q-Bernoulli and Euler polynomials, Advances in Difference Equations 2013, 2013:108.

8

# ON THE BOUNDS OF THE FIRST REFORMULATED ZAGREB INDEX

**T. Mansour[1], M. A. Rostami[2], E. Suresh[3,*], G. B. A. Xavier[4]**

[1]Department of Mathematics, University of Haifa, 3498838 Haifa, Israel

[2]Institute for Computer Science, Friedrich Schiller University Jena, Germany

[3]Department of Mathematics, Velammal Engineering College, Surapet, Chennai-66, Tamil Nadu, India

[4]Department of Mathematics, Sacred Heart College, Tirupattur-635601, Tamil Nadu, India

## ABSTRACT

The edge version of traditional first Zagreb index is known as first reformulated Zagreb index. In this paper, we analyze and compare various lower and upper bounds for the first reformulated Zagreb index and we propose new lower and upper bounds which are stronger than the existing and recent results [*Appl. Math. Comp.* **273** (2016) 16-20]. In addition, we prove that our bounds are superior in comparison with the other existing bounds.

## KEYWORDS

Zagreb indices, first reformulated Zagreb index, forgotten topological index

## 1 INTRODUCTION

Throughout this paper, we consider finite connected undirected simple graphs. Let $G$ be such a graph with $n$ vertices and $m$ edges. The degree of a vertex $u$ is denoted by $d(u)$. Especially, $\Delta = \Delta(G)$ and $\delta = \delta(G)$ are called as maximum and minimum degree of $G$, espectively. The degree of an edge is denoted by $d(e)$ in $G$, which is defined by $d(e) = d(u)+d(v)-2$ for $e = uv \in E(G)$. Let $\bar{G}$ denotes the complement of $G$, with the same vertex set such that two vertices $u$ and $v$ are adjacent in $\bar{G}$ if and only if they are not adjacent in $G$.

The first and second Zagreb indices are defined as (for instance, see [9] and references cited therein)

$$M_1^2(G) = \sum_{v \in V(G)} d(v)^2$$

$$\text{and } M_2^1(G) = \sum_{uv \in E(G)} d(u)d(v).$$

Numerous papers were recorded in the literature regarding the mathematical and chemical properties on Zagreb indices and the surveys on Zagreb indices. For the recent outcomes on Zagreb indices, see [5] and references cited therein. In 2005, Li and Zheng [13] introduced the the generalized version of the first Zagreb index. For $\alpha \in \mathbb{R}$ and $G$ be any graph, it satisfies

$$M_1^{\alpha+1}(G)=\sum_{v\in V(G)} d(v)^{\alpha+1}=\sum_{uv\in E(G)}\left[d(u)^{\alpha}+d(v)^{\alpha}\right] \tag{1.1}$$

In 2015, Fortula and Gutman [8] re-introduced the forgotten topological index:

$$F(G)=M_1^3(G)=\sum_{v\in V(G)} d(v)^3.$$

Mathematical properties of the forgotten topological index is seen in [4, 8, 11, 19]. Miličević, Nikolić and Trinajstić [15] reformulated the first Zagreb index in terms of edge-degree instead of vertex-degree:

$$EM_1^2(G)=\sum_{e\in E(G)} d(e)^2.$$

The first reformulated Zagreb index is simply the Zagreb index of its line graph $L(G)$. The use of these descriptors in QSPR study was also discussed in their report [15]. In [6, 11, 12, 17, 20], various properties, relations and bounds are also explored.

The inverse degree was first appeared through conjectures of the computer program Graffiti [7]denote by $ID(G)$; with no isolated vertices defined by

$$ID(G)=\sum_{v\in V(G)} \frac{1}{d(v)}.$$

The inverse degree is a special case formula at $\alpha=-1$ in (1.1). For the recent results of the inverse degree, see [3, 18]. In analogy, the inverse edge degree of a graph $G$ with no isolated edges is defined by

$$IED(G)=\sum_{e\in E(G)} \frac{1}{d(e)}.$$

This paper is organized as follows. In Section 2, we give some preliminaries and basic definitions that will be used throughout the paper. In Sections 3 and 4, we present lower and upper bounds on the first reformulated Zagreb index $EM_1^2(G)$ in terms of $n, m, \Delta, \delta, ID(G)$, $IED(G)$, $M_1^2(G)$ and $M_2^1(G)$ respectively. Furthermore, (Refer Table 3-Table 6) we showed that our bounds have the smallest deviation from the first reformulated Zagreb index $EM_1^2(G)$ and superior than the existing bounds in the literature.

# 2 PRELIMINARIES

As usual $P_n, K_{1,n-1}$ and $C_n$ denote a path, star and a cycle of order *n*, respectively. Let *G* and *H* be graphs. We denote the number of distinct subgraphs of the graph *G* which are isomorphic to *H* by $\sigma_G(H)$. In specific, *G* has $\sigma_G(P_1)$ vertices, $\sigma_G(P_2)$ edges and $\sigma_G(C_3)$ triangles. Let $\alpha$ and $\beta$ be positive integers, then the double star $D_{\alpha,\beta}$ is a tree on $\alpha+\beta+2$ vertices obtained from the path $P_2$, by attaching $\alpha$ pendent vertices to its one vertex, and $\beta$ pendent vertices to its other vertex.

GraphTea [1] is a software tool for graph problems with a focus on extracting information and visualization. It offers powerful ways to query or directly interact with properties of a particular instance of a graph problem. It is specially designed for the computational works on the topological indices.

To analyse and compare our results, we need to recall few classes of graphs:

- The *wheel* $W_n$ is join of the graphs $C_{n-1}$ and $K_1$.
- The *crown* $Cr_n$ is obtained from the cycle $C_n$ by adjoining a pendant edge at each vertex of the cycle.

- The *helm* $H_n$ is obtained from the wheel graph $W_{n-1}$ by adjoining a pendant edge at each vertex of the cycle.
- The *flower* $Fl_n$ is obtained from the helm $H_n$ by joining each pendent vertex to the central vertex of the helm.
- A graph $G$ is called *bidegreed* if its vertex degree is either $\Delta$ or $\delta$ with $\Delta > \delta \geq 1$.

In 2010, Zhou and Trinajstić [20] obtained the following identity.

**Proposition 2.1.** *Let G be a simple graph with n vertices and m edges. Then*

$$EM_1^2(G) = M_1^3(G) + 2M_2^1(G) - 4M_1^2(G) + 4m. \tag{2.1}$$

Using the results from [4] without exaggeration, we obtain the following corollary for $EM_1^2(G)$ in terms of its subgraphs of *G*.

**Corollary 2.2.** *Let G be a simple graph with n vertices and m edges. Then*

$$EM_1^2(G) = 6\sigma_G(K_{1,3}) + 2\sigma_G(P_4) + 2\sigma_G(P_3) + 6\sigma_G(P_3).$$

# 3 LOWER BOUNDS ON THE FIRST REFORMULATED ZAGREB INDEX

In 2012, Ilić and Zhou [11] proposed the following lower bound for $EM_1^2(G)$.

**Lemma 3.1.** *Let G be a simple graph with n vertices and m edges. Then*

$$EM_1^2(G) \geq \frac{2m}{n} M_1^2(G) + 2M_2^1(G) - 4M_1^2(G) + 4m \tag{3.1}$$

*with equality if and only if G is regular.*

Very recently in 2016, Milovanović, Dolićanin, Glogić [16] obtained the new lower bounds for $EM_1^2(G)$ in terms of $M_1^2(G)$ and $M_2^1(G)$.

**Lemma 3.2.** *Let G be a simple graph with* $n(n \geq 2)$ *vertices and m edges. Then*

$$EM_1^2(G) \geq \frac{M_1^2(G)^2}{2m} + 2M_2^1(G) - 4M_1^2(G) + 4m \tag{3.2}$$

$$EM_1^2(G) \geq 4\left(M_2^1(G) - M_1^2(G) + m\right)$$

(3.3)*equality holds if and only if G is a regular.*

**Remark 3.3.** *In 2015, Fortula and Gutman* [8] *have given the following lower bounds for* $M_1^3(G)$. *Let G be a graph with m edges. Then*

$$M_1^3(G) \geq \frac{\left(M_1^2(G)\right)^2}{2m} \tag{3.4}$$

$$M_1^3(G) \geq \frac{\left(M_1^2(G)\right)^2}{2m} - 2M_2^1(G). \tag{3.5}$$

*Note that, the equality of* (3.4) *holds if and only if G is regular. It is easy to see that, the inequality*(3.2) *can be obtained from* (2.1) *by using* (3.4).

In analogy, one can propose a lower bound for $EM_1^2(G)$ using (3.5). But in 2012, De [6] obtained the following inequality.

**Lemma 3.4.** *Let G be a simple graph with* $n(n \geq 2)$ *vertices and m edges. Then*

$$EM_1^2(G) \geq \frac{\left(M_1^2(G)\right)^2}{m} - 4M_1^2(G) + 4m \tag{3.6}$$

*equality holds if and only if G is a regular.*

In 2010, Zhou and Trinajstić [19] proposed the following inequality in the context of general sum connectivity.

**Lemma 3.5.** *Let G be a graph with* $m \geq 1$ *edges. If* $0 < \alpha < 1$, *then* $\chi^{\alpha}(G) \leq M_1^2(G)^{\alpha} m^{1-\alpha}$, *and if* $\alpha < 0$ *or* $\alpha > 1$, *then* $\chi^{\alpha}(G) \geq M_1^2(G)^{\alpha} m^{1-\alpha}$, *and either equality holds if and only if d(u) + d(v) is a constant for any edge uv.*

Note that, for $\alpha$ = 2 in Lemma 3.5 turns (3.5) as its special case. Also from Lemma 3.5, it is clear that the equality of (3.6) holds if and only if *d*(*u*) + *d*(*v*) is a constant for any edge *uv:* In addition, from [8], it is easy to see that (3.2) and (3.6) are incomparable.

Now, we are ready to present our lower bounds on $EM_1^2(G)$ in terms of $n, m, \Delta, \delta, ID(G), M_1^2(G)$ and $M_2^1(G)$.

**Theorem 3.6.** *Let G be a simple graph with n vertices and m edges, then*

$$EM_1^2(G) \geq \frac{M_1^2(G)^2}{2m} + \frac{1}{2m}\left(2mID(G) - n^2\right) + 2M_2^1(G) - 4M_1^2(G) + 4m \tag{3.7}$$

*equality holds if and only if G is regular.*

*Proof.* Consider $w_1, w_2, \ldots, w_k$ be the non-negative weights, then we have the weighted version of Cauchy-Schwartz inequality

$$\sum_{i=1}^{k} w_i a_i^2 \sum_{i=1}^{k} w_i b_i^2 \geq \left(\sum_{i=1}^{k} w_i a_i b_i\right)^2. \tag{3.8}$$

Since $w_i$ is non-negative, we assume that $w_i = m_i - n_i$ such that $m_i \geq n_i \geq 0$. Thus

$$\sum_{i=1}^{k} m_i a_i^2 \sum_{i=1}^{k} m_i b_i^2 - \left(\sum_{i=1}^{k} m_i a_i b_i\right)^2$$
$$\geq \sum_{i=1}^{k} n_i a_i^2 \sum_{i=1}^{k} n_i b_i^2 - \left(\sum_{i=1}^{k} n_i a_i b_i\right)^2 \geq 0. \quad (3.9)$$

Set $m_i = d(v_i), n_i = \frac{1}{d(v_i)}, a_i = d(v_i)$ and $b_i = 1,$ for all $i = 1, 2, \ldots, n$ in the above, we get

$$\sum_{i=1}^{n} d(v_i)^3 \sum_{i=1}^{n} d(v_i) - \left(\sum_{i=1}^{n} d(v_i)^2\right)^2$$
$$\geq \sum_{i=1}^{n} d(v_i) \sum_{i=1}^{n} \frac{1}{d(v_i)} - \left(\sum_{i=1}^{n} 1\right)^2.$$

The inequality (3.7) is obtained from the above inequality and equality (2.1) and the equality holds if and only if *G* is regular.

**Theorem 3.7.** *Let G be a simple graph with n vertices and m edges, then*

$$EM_1^2(G) \geq \frac{M_1^2(G)^2}{2m} + \frac{1}{2m}\left(nM_1^2(G) - 4m^2\right)$$
$$+ 2M_2^1(G) - 4M_1^2(G) + 4m \quad (3.10)$$

*equality holds if and only if G is regular.*

*Proof.* The proof follows from the same terminology of Theorem 3.6 by choosing $m_i = d(v_i),\ n_i = 1, a_i = d(v_i)$ and $b_i = 1,$ for all $i = 1, 2, \ldots, n.$

**Remark 3.8.** For every simple graph *G*, the lower bound in (3.10) is always better than the lower bound in (3.7). For this, we have to show that

$$nM_1^2(G)-4m^2 \geq 2mID(G)-n^2 \geq 0. \tag{3.11}$$

*By fixing* $a_i = d(v_i), b_i = 1, m_i = 1$ *and* $n_i = d(v_i)^{-1}$ *in* (3.9), *we achieve our required claim and the equality holds if and only if G is regular. Therefore inequality* (3.10) *is stronger than* (3.7) *and* (3.7)*is stronger than* (3.2).

**Remark 3.9.** *On the other hand, the lower bounds in* (3.6) *and* (3.10) *are incomparable. For Helm* $H_n(n\geq 4)$ *the lower bound in* (3.10) *is better than* (3.6) *and for the Flower* $F_n(n\geq 4)$ *the lower bound in* (3.6) *is better than* (3.10). *In addition for the Crown* $Cr_n$ *both the lower bounds coincides together, other than the equality case.*

Next, we need to improve the lower bound (3.6). For which we need the following inequality which relates the first Zagreb index, second Zagreb index and the forgotten topological index.

**Theorem 3.10.** *Let G be a simple graph with no isolated edges, then*

$$M_1^3(G) \geq \frac{M_1^2(G)^2}{m} + \frac{1}{m}\left[M_1^2(G)-2m\right]IED(G) - 2M_2^1(G)-m \tag{3.12}$$

*equality holds if and only if* $d(u) + d(v)$ *is a constant for any edge uv.*

*Proof.* Considering the weighted version of Cauchy-Schwartz inequality from (3.8) with the assumption $w_i = x_i - y_i$ such that $x_i \geq y_i \geq 0$ and by setting $x_i = 1$, $y_i = \frac{1}{d(e_i)}, a_i = d(e_i)$ and $b_i = 1$, for all $i = 1, 2, \ldots, m$, we get

$$\sum_{i=1}^{m} d(e_i)^2 \sum_{i=1}^{m} 1 - \left(\sum_{i=1}^{m} d(e_i)\right)^2 \geq \sum_{i=1}^{m} d(e_i) \sum_{i=1}^{m} \frac{1}{d(e_i)} - \left(\sum_{i=1}^{m} 1\right)^2$$

expanding, we have

$$\begin{pmatrix} M_1^3(G)+M_2^1(G) \\ +4m-4M_1^2(G) \end{pmatrix} m-\left(M_1^2(G)-2m\right)^2$$
$$\geq\left(M_1^2(G)-2m\right)IED(G)-m^2,$$

and the equality holds if and only if $d(u) + d(v)$ is a constant for any edge $uv$, which completes our claim.

**Corollary 3.11.** *With the assumptions in Theorem* 3.10, *we have*

$$EM_1^2(G)\geq\frac{M_1^2(G)^2}{m}+\frac{1}{m}\left[M_1^2(G)-2m\right]IED(G)$$
$$-4M_1^2(G)+3m \tag{3.13}$$

*equality holds if and only if* $d(u) + d(v)$ *is a constant for any edge* $uv$.

**Remark 3.12.** *Still the lower bound in* (3.10) *and* (3.13) *are incomparable. Consider the Helm* $H_n(n\geq 4)$, *for which the lower bound in* (3.10) *is finer than* (3.13) *and for the complement* $H_n(n\geq 4)$ *the lower bound in* (3.13) *is finer than* (3.10).

Our next interest is to reduce the deviation for the lower bounds from the first reformulated Zagreb index. To achieve our goal, either we need to fit some good lower bounds for $M_1^3$ or some good upper bounds for $M_1^2(G)$ in the equality (2.1).

In 2011, Ilić and Liu [10] proposed a new upper bound for the first Zagreb index $M_1^2(G)$ and proved that it has the smallest deviation from the first Zagreb index.

**Lemma 3.13.** *Let G be a simple regular graph with n vertices and m edges, with a vertices of maximum degree* $\Delta$ *and b vertices of minimum degree* $\delta$. *Then*

$$M_1^2(G)\leq 2m(\Delta+\delta)-n\Delta\delta-(\Delta-\delta-1)(n-a-b), \tag{3.14}$$

*equality holds if and only if the vertex degrees are equal to* $\delta$, $\delta+1$, $\Delta-1$ *or* $\Delta$.

Thus we can state the following result.

**Corollary 3.14.** *With the assumptions in Lemma 3.13, we have*

$$EM_1^2(G) \geq M_1^3(G) + 2M_2^1(G) - 8m(\Delta+\delta) + 4n\Delta\delta \\ + 4(\Delta-\delta-1)(n-a-b) + 4m, \tag{3.15}$$

To the best knowledge of the authors, still now in the literature (3.14) is superior upper bound for the first Zagreb index $M_1^2(G)$. This motivates us to present the new upper bound which is stronger than (3.14) and which leads to be the better lower bound for the first reformulated Zagreb index $EM_1^2(G)$.

**Theorem 3.15.** *Let G be a simple graph with n non-isolated vertices and m edges, with a vertices of maximum degree* $\Delta$ *and b vertices of minimum degree* $\delta$. *Then*

$$M_1^2(G) \leq a\Delta^2 + b\delta^2 + (2m - a\Delta - b\delta)(\Delta+\delta+1) \\ + (\Delta-1)(\delta+1)\left[ID(G) - \frac{a}{\Delta} - \frac{b}{\delta}\right] \\ - (\Delta\delta - 2\Delta - 1)(n-a-b), \tag{3.16}$$

*equality holds if and only if the vertex degrees are equal to* $\delta$, $\delta+1$, $\Delta-1$ *or* $\Delta$.

*Proof.* Let $a, A \in \mathbb{R}$ and $x_i, y_i$ be two sequences with the property $ay_i \leq x_i \leq Ay_i$ for $i = 1, 2, \ldots, n$ and $w_i$ be any sequence of positive real numbers, it holds

Since $w_i$ is a positive sequence, choose $w_i = m_i - n_i$ such that $m_i \geq n_i$ and by setting $A = \Delta$, $a = \delta$, $x_i = d(v_i)$, $y_i = 1$, $m_i = 1$ and $n_i = \frac{1}{d(v_i)}$ and adding over the vertices by restricting $\delta < d(v_i) < \Delta$, we have

$$\sum_{\delta<d(v_i)<\Delta}\left(1-\frac{1}{d(v_i)}\right)(\Delta-d(v_i))(d(v_i)-\delta)$$
$$\geq \sum_{\delta<d(v_i)<\Delta}\left(1-\frac{1}{d(v_i)}\right)(\Delta-\delta-1)\geq 0, \tag{3.17}$$

expanding the above inequality gives the required result with equality if and only if $d(v_i)=\delta+1$ or $d(v_i)=\Delta-1$. This completes the proof.

The computational results for connected graphs on $n$ = 3 to 9 vertices and connected trees on $n$ = 10 to 20 vertices are provided in Table 1 and Table 2 respectively.

In the group one of the Table 1 and Table 2, the first column represents the degree of the vertex $n$,the second column contains number of connected graphs (trees) on $n$ vertices and the third one has the average value of the first Zagreb index $M_1^2(G)$. Next two groups of three columns represent the average value of the upper bound, the standard deviation

$$\left(\sqrt{\frac{\sum_G\left(X(G)-M_1^2(G)\right)^2}{count}}\right)$$

and the number of graphs with equality case.

**Table 1.** Upper bound comparision of $M_1^2(G)$ for small graphs

| | Parameters | | Parameters | | | Lemma 3.13 | | |
|---|---|---|---|---|---|---|---|---|
| n | Count | Avg. | Avg. | Stdev. | Eq. | Avg. | Stdev. | Eq. |
| 3 | 2 | 9.0000 | 9.0000 | 0.0000 | 2 | 9.0000 | 0.0000 | 2 |
| 4 | 6 | 19.6667 | 19.6667 | 0.0000 | 6 | 19.6667 | 0.0000 | 6 |
| 5 | 21 | 35.4286 | 35.4286 | 0.0000 | 21 | 35.4286 | 0.0000 | 21 |
| 6 | 112 | 55.6607 | 55.7440 | 0.3883 | 106 | 55.7857 | 0.5825 | 106 |
| 7 | 853 | 82.6260 | 83.0240 | 1.1093 | 683 | 83.1981 | 1.5869 | 683 |
| 8 | 11117 | 118.4507 | 119.6036 | 2.4343 | 6658 | 120.0452 | 3.3453 | 6658 |
| 9 | 261080 | 166.1056 | 168.6350 | 4.5110 | 105659 | 169.4830 | 5.9966 | 105659 |

**Table 2.** Upper bound comparison of $M_1^2(G)$ for small trees

| | Parameters | | Parameters | | | Lemma 3.13 | | |
|---|---|---|---|---|---|---|---|---|
| n | Count | Avg. | Avg. | Stdev. | Eq. | Avg. | Stdev. | Eq. |
| 10 | 106 | 44.5849 | 44.7311 | 0.4200 | 91 | 44.8019 | 0.6219 | 91 |
| 11 | 235 | 50.0255 | 50.2645 | 0.6064 | 189 | 50.3787 | 0.8921 | 189 |
| 12 | 551 | 55.4011 | 55.7551 | 0.8162 | 413 | 55.9220 | 1.1936 | 413 |
| 13 | 1301 | 60.7640 | 61.2498 | 1.0509 | 913 | 61.4766 | 1.5299 | 913 |
| 14 | 3159 | 66.1292 | 66.7626 | 1.3027 | 2075 | 67.0560 | 1.8898 | 2075 |
| 15 | 7741 | 71.4949 | 72.2915 | 1.5748 | 4774 | 72.6583 | 2.2783 | 4774 |
| 16 | 19320 | 76.8598 | 77.8323 | 1.8611 | 11214 | 78.2780 | 2.6862 | 11214 |
| 17 | 48629 | 82.2303 | 83.3931 | 2.1644 | 26619 | 83.9238 | 3.1180 | 26619 |
| 18 | 123867 | 87.6032 | 88.9683 | 2.4803 | 64057 | 89.5893 | 3.5673 | 64057 |
| 19 | 317955 | 92.9794 | 94.5599 | 2.8104 | 155575 | 95.2768 | 4.0365 | 155575 |
| 20 | 823065 | 98.3585 | 100.1660 | 3.1522 | 381521 | 100.9836 | 4.5219 | 381521 |

On comparison of the computational results in Table 1 and Table 2, we observe it is coherent that the upper bound (3.16) has the least deviation from $M_1^2(G)$ and stronger than (3.14).

**Corollary 3.16.** *With the assumptions in Theorem 3.15, we have*

$$\begin{aligned} EM_1^2(G) \geq\ & M_1^3(G) + 2M_2^1(G) \\ & + 4(\Delta - \delta - 1)(n - a - b) + 4m \\ & - 4(\Delta - 1)(\delta + 1)\left[ ID(G) - \frac{a}{\Delta} - \frac{b}{\delta} \right] \\ & - 4\left(a\Delta^2 + b\delta^2\right) \\ & - 4(2m - a\Delta - b\delta)(\Delta + \delta + 1) \end{aligned} \tag{3.18}$$

*equality holds if and only if the vertex degrees are equal to* $\delta$, $\delta+1$, $\Delta-1$ *or* $\Delta$.

In analogy, Table 3 - Table 4 conclude that the lower bound (3.18) has the least deviation from $M_1^2(G)$ and so it is superior than the existing lower bounds.

# 4 UPPER BOUNDS ON THE FIRST REFORMULATED ZAGREB INDEX

In 2010, Zhou and Trinajstić [20] proposed the first upper bound for $EM_1^2(G)$.

**Lemma 4.1.** *Let G be a graph on n vertices and* $m \geq 1$ *edges. Then*

$$EM_1^2(G) \leq (n-4)M_1^2(G) + 4M_2^1(G) - 4m^2 + 4m \tag{4.1}$$

*equality holds if and only if any two non-adjacent vertices have equal degrees.*

**Table 3.** Lower bound comparison of $EM_1^2(G)$ for small graphs

| | Parameters | | Corollary 3.16 | | | Corollary 3.14 | | |
|---|---|---|---|---|---|---|---|---|
| n | Count | Avg. | Avg. | Stdev. | Eq. | Avg. | Stdev. | Eq. |
| 3 | 2 | 7.0000 | 7.0000 | 0.0000 | 2 | 7.0000 | 0.0000 | 2 |
| 4 | 6 | 34.6667 | 34.6667 | 0.0000 | 6 | 34.6667 | 0.0000 | 6 |
| 5 | 21 | 96.2857 | 96.2857 | 0.0000 | 21 | 96.2857 | 0.0000 | 21 |
| 6 | 112 | 195.9643 | 195.6310 | 1.5533 | 106 | 195.4643 | 2.3299 | 106 |
| 7 | 853 | 356.3329 | 354.7409 | 4.4371 | 683 | 354.0445 | 6.3475 | 683 |
| 8 | 11117 | 605.4936 | 600.8817 | 9.7374 | 6658 | 599.1156 | 13.3811 | 6658 |
| 9 | 261080 | 991.0099 | 980.8922 | 18.0439 | 105659 | 977.5002 | 23.9863 | 105659 |
| | Parameters | | Theorem 3.7 | | | Corollary 3.11 | | |
| n | Count | Avg. | Avg. | Stdev. | Eq. | Avg. | Stdev. | Eq. |
| 3 | 2 | 7.0000 | 6.7500 | 0.3536 | 1 | 7.0000 | 0.0000 | 2 |
| 4 | 6 | 34.6667 | 33.1389 | 2.1093 | 2 | 34.3500 | 0.4732 | 3 |
| 5 | 21 | 96.2857 | 91.2316 | 6.4066 | 2 | 94.3278 | 2.3736 | 4 |
| 6 | 112 | 195.9643 | 185.4258 | 13.0763 | 5 | 190.4460 | 6.3778 | 7 |
| 7 | 853 | 356.3329 | 336.8463 | 23.2882 | 4 | 344.2109 | 13.6459 | 7 |
| 8 | 11117 | 605.4936 | 573.8401 | 36.7729 | 17 | 583.5165 | 24.4070 | 20 |
| 9 | 261080 | 991.0099 | 942.5787 | 54.9013 | 22 | 954.9776 | 39.5418 | 27 |

**Table 4.** Lower bound comparison of $EM_1^2(G)$ for small trees

| | Parameters | | Corollary 3.16 | | | Corollary 3.14 | | |
|---|---|---|---|---|---|---|---|---|
| *n* | Count | Avg. | Avg. | Stdev. | Eq. | Avg. | Stdev. | Eq. |
| 10 | 106 | 99.4151 | 98.8302 | 1.6798 | 91 | 98.5472 | 2.4877 | 91 |
| 11 | 235 | 114.4340 | 113.4780 | 2.4256 | 189 | 113.0213 | 3.5682 | 189 |
| 12 | 551 | 128.7477 | 127.3316 | 3.2649 | 413 | 126.6642 | 4.7744 | 413 |
| 13 | 1301 | 142.8932 | 140.9500 | 4.2035 | 913 | 140.0430 | 6.1195 | 913 |
| 14 | 3159 | 156.9839 | 154.4501 | 5.2108 | 2075 | 153.2764 | 7.5592 | 2075 |
| 15 | 7741 | 171.0559 | 167.8697 | 6.2993 | 4774 | 166.4023 | 9.1131 | 4774 |
| 16 | 19320 | 185.1062 | 181.2163 | 7.4443 | 11214 | 179.4335 | 10.7449 | 11214 |
| 17 | 48629 | 199.2024 | 194.5513 | 8.6574 | 26619 | 192.4283 | 12.4720 | 26619 |
| 18 | 123867 | 213.3222 | 207.8617 | 9.9211 | 64057 | 205.3778 | 14.2694 | 64057 |
| 19 | 317955 | 227.4755 | 221.1534 | 11.2414 | 155575 | 218.2861 | 16.1458 | 155575 |
| 20 | 823065 | 241.6596 | 234.4294 | 12.6088 | 381521 | 231.1590 | 18.0876 | 381521 |
| | Parameters | | Theorem 3.7 | | | Corollary 3.11 | | |
| *n* | Count | Avg. | Avg. | Stdev. | Eq. | Avg. | Stdev. | Eq. |
| 10 | 106 | 99.4151 | 71.2432 | 45.7847 | 0 | 88.7336 | 13.3439 | 1 |
| 11 | 235 | 114.4340 | 82.1421 | 52.2828 | 0 | 101.0084 | 17.0481 | 1 |
| 12 | 551 | 128.7477 | 92.8494 | 57.5453 | 0 | 112.6745 | 20.6571 | 1 |
| 13 | 1301 | 142.8932 | 103.3729 | 62.4065 | 0 | 124.1590 | 24.3165 | 1 |
| 14 | 3159 | 156.9839 | 113.9656 | 66.7328 | 0 | 135.6441 | 27.8509 | 1 |
| 15 | 7741 | 171.0559 | 124.5123 | 70.8548 | 0 | 147.1143 | 31.3411 | 1 |
| 16 | 19320 | 185.1062 | 135.0994 | 74.7407 | 0 | 158.6074 | 34.7026 | 1 |
| 17 | 48629 | 199.2024 | 145.7020 | 78.5596 | 0 | 170.1444 | 38.0023 | 1 |
| 18 | 123867 | 213.3222 | 156.3424 | 82.3052 | 0 | 181.7243 | 41.2101 | 1 |
| 19 | 317955 | 227.4755 | 167.0016 | 86.0265 | 0 | 193.3383 | 44.3579 | 1 |
| 20 | 823065 | 241.6596 | 177.6906 | 89.7288 | 0 | 204.9893 | 47.4428 | 1 |

In 2012, Ilić and Zhou [11] obtained the following upper bound for $EM_1^2(G)$.

**Lemma 4.2.** *Let G be a simple graph with n vertices and m edges. Then*

$$EM_1^2(G) \le (\Delta + \delta - 4) M_1^2(G) + 2M_2^1(G) - 2m\Delta\delta + 4m \tag{4.2}$$

*equality holds if and only if G is a regular or bidegreed graph.*

In the same time period, De [6] Proposed the following upper bounds for $EM_1^2(G)$.

**Lemma 4.3.** *Let G be a simple graph with n vertices and m edges. Then*

$$EM_1^2(G) \le 2(\Delta + \delta - 2) M_1^2(G) - 4m(\Delta\delta - 1) \tag{4.3}$$

$$EM_1^2(G) \le \left(M_1^2(G) - 2m\right)(m + 2\delta - 3) - 2m(m-1)(\delta - 1) \tag{4.4}$$

*equality holds if and only if G is regular.*

In 2016, Milovanović et. al. [16] obtained some new upper bounds for the first reformulated Zagreb index.

**Lemma 4.4.** *Let G be a simple graph with* $n(n \ge 2)$ *vertices and m edges. Then*

$$EM_1^2(G) \le \frac{M_1^2(G)^2}{2m} + 2M_2^1(G) - 4M_1^2(G) + \frac{m}{2}(\Delta - \delta)^2 + 4m \tag{4.5}$$

*equality holds if and only if G is regular or* $G \cong K_{1,n-1}$. *Moreover,*

$$EM_1^2(G) \leq \frac{\left(M_1^2(G) - 2m\right)^2}{m} + m(\Delta - \delta)^2 \tag{4.6}$$

$$EM_1^2(G) \leq \frac{\left(M_1^2(G) - 2m\right)^2}{m} + 2m\left(M_1^2(G) - \frac{4m^2}{n}\right) \tag{4.7}$$

*equality holds if and only if G is regular.*

**Remark 4.5.** *The upper bounds in* (4.2) *and* (4.3) *are emerged from Diaz-metcalf inequality, by considering the vertex and edge degrees respectively. At first, it is easy to see that the bounds in*(4.1) *and* (4.2) *are always better than* (4.3) *and* (4.4). *Secondly, the upper bound* (4.2) *is always better than the upper bounds* (4.5), (4.6) *and* (4.7), *so we leave it to the fascinated reader. In addition, the bounds in* (4.1) *and* (4.2) *are incomparable. For Helm Graphs* $H_n$, (4.2) *is better than*(4.1), *and for the Flower graphs* $F_n$, (4.1) *is better than* (4.2).

Next, we are ready to present our inequality that determines upper bound for the forgotten topological index $M_1^3(G)$ in terms of $n, m, \Delta, M_1^2(G)$, and $M_2^1(G)$, which leads to the upper bound for $EM_1^2(G)$.

**Theorem 4.6.** *Let G be a simple graph with n non-isolated vertices and m edges, with a vertices of maximum degree* $\Delta$ *and b vertices of minimum degree* $\delta$. *Then*

$$M_1^3(G) \le a\Delta^3 + b\delta^3 + (\Delta+\delta)\begin{pmatrix} M_1^2(G) - a(\Delta^2+1) \\ -b(\delta^2+1) - n \end{pmatrix}$$
$$+(\Delta-1)(\delta+1)\left[ID(G) - \frac{a}{\Delta} - \frac{b}{\delta}\right]$$
$$-((\Delta-1)(\delta+1)-1)(2m - a\Delta - b\delta) \qquad (4.8)$$

*equality holds if and only if the vertex degrees are equal to* $\delta$, $\delta+1$, $\Delta-1$ *or* $\Delta$.

*Proof.* The proof follows the same terminology of Theorem 3.15, by the choice of setting $m_i = d(v_i)$ and $n_i = \frac{1}{d(v_i)}$.

**Corollary 4.7.** *With the assumptions in Theorem 4.6, we have*

$$EM_1^2(G)$$
$$\le a\Delta^3 + b\delta^3 + (\Delta-1)(\delta+1)\left[ID(G) - \frac{a}{\Delta} - \frac{b}{\delta}\right]$$
$$+(\Delta-\delta)\left(M_1^2(G) - a(\Delta^2+1) - b(\delta^2+1) - n\right)$$
$$+2M_2^1(G) - ((\Delta-1)(\delta+1)-1)(2m - a\Delta - b\delta)$$
$$-4M_1^2(G) + 4m \qquad (4.9)$$

*equality holds if and only if the vertex degrees are equal to* $\delta$, $\delta+1$, $\Delta-1$ *or* $\Delta$.

**Corollary 4.8.** *Let G be a simple graph with n vertices and m edges, with a vertices of maximum degree* $\Delta$ *and b vertices of minimum degree* $\delta$. *Then*

$$\begin{aligned}
&EM_1^2(G)\\
&\leq a\Delta^3 + b\delta^3 + (\Delta+\delta+1)\left(M_1^2(G) - a\Delta^2 - b\delta^2\right)\\
&+(\Delta-1)(\delta+1)(n-a-b)\\
&-(\Delta\delta+2\Delta-1)(2m-a\Delta-b\delta)\\
&+2M_2^1(G) - 4M_1^2(G) + 4m
\end{aligned} \tag{4.10}$$

*equality holds if and only if the vertex degrees are equal to* $\delta$, $\delta+1$, $\Delta-1$ *or* $\Delta$.

*Proof.* The proof follows the same terminology of Theorem 3.15, by the choice of setting $m_i = d(v_i)$ and $n_i = 1$ and using the equality (2.1).

From Table 5-Table 6 (with same notations in Table 1-Table 2), our upper bound (4.10) has the small deviation from $M_1^2(G)$ which concludes that (4.10) is the superior than the existing upper bounds.

We end the paper by remarking that the lower and upper bounds of $EM_1^2(G)$ is determined in terms of *n; m;* $ID(G), M_1^2(G)$ and $M_2^1(G)$. In addition, we presented the upper bound for the first Zagreb index $M_1^2(G)$ and the forgotten topological index $M_1^3(G)$ and with the assistance of computational results, we conclude that these inequalities are superior than the other bounds appeared in the literature so far.

**Table 5.** Upper bound comparison of $EM_1^2(G)$ for small graphs

| | Parameters | | Corollary 4.8 | | | Corollary 4.7 | | |
|---|---|---|---|---|---|---|---|---|
| *n* | Count | Avg. | Avg. | Stdev. | Eq. | Avg. | Stdev. | Eq. |
| 3 | 2 | 7.0000 | 7.0000 | 0.0000 | 2 | 7.0000 | 0.0000 | 2 |
| 4 | 6 | 34.6667 | 34.6667 | 0.0000 | 6 | 34.6667 | 0.0000 | 6 |
| 5 | 21 | 96.2857 | 96.2857 | 0.0000 | 21 | 96.2857 | 0.0000 | 21 |
| 6 | 112 | 195.9643 | 196.2143 | 1.1650 | 106 | 196.2976 | 1.5533 | 106 |
| 7 | 853 | 356.3329 | 357.6764 | 3.8272 | 683 | 358.0744 | 4.9305 | 683 |
| 8 | 11117 | 605.4936 | 609.8366 | 9.4456 | 6658 | 610.9895 | 11.8628 | 6658 |
| 9 | 261080 | 991.0099 | 1001.5845 | 19.3717 | 105659 | 1004.1139 | 23.8496 | 105659 |
| | Parameters | | Lemma 4.1 | | | Lemma 4.2 | | |
| *n* | Count | Avg. | Avg. | Stdev. | Eq. | Avg. | Stdev. | Eq. |
| 3 | 2 | 7.0000 | 7.0000 | 0.0000 | 2 | 7.0000 | 0.0000 | 2 |
| 4 | 6 | 34.6667 | 35.3333 | 1.6330 | 5 | 35.3333 | 1.1547 | 4 |
| 5 | 21 | 96.2857 | 99.6190 | 5.9362 | 11 | 99.6190 | 4.9570 | 8 |
| 6 | 112 | 195.9643 | 205.9107 | 15.0772 | 36 | 206.0000 | 13.0767 | 15 |
| 7 | 853 | 356.3329 | 380.2673 | 31.8992 | 91 | 378.1290 | 26.5012 | 25 |
| 8 | 11117 | 605.4936 | 652.0770 | 58.6369 | 471 | 644.4005 | 45.5048 | 53 |
| 9 | 261080 | 991.0099 | 1072.0948 | 98.1035 | 2296 | 1052.3611 | 70.0854 | 95 |

**Table 6.** Upper bound comparison of $EM_1^2(G)$ for small trees

| | Parameters | | Corollary 4.8 | | | Corollary 4.7 | | |
|---|---|---|---|---|---|---|---|---|
| *n* | Count | Avg. | Avg. | Stdev. | Eq. | Avg. | Stdev. | Eq. |
| 10 | 106 | 99.4151 | 99.8679 | 1.3175 | 91 | 100.0142 | 1.7358 | 91 |
| 11 | 235 | 114.4340 | 115.1830 | 1.9439 | 189 | 115.4220 | 2.5469 | 189 |
| 12 | 551 | 128.7477 | 129.8730 | 2.6727 | 413 | 130.2270 | 3.4831 | 413 |
| 13 | 1301 | 142.8932 | 144.4520 | 3.4937 | 913 | 144.9377 | 4.5362 | 913 |
| 14 | 3159 | 156.9839 | 159.0326 | 4.3828 | 2075 | 159.6660 | 5.6745 | 2075 |
| 15 | 7741 | 171.0559 | 173.6478 | 5.3461 | 4774 | 174.4444 | 6.9074 | 4774 |
| 16 | 19320 | 185.1062 | 188.2864 | 6.3635 | 11214 | 189.2589 | 8.2089 | 11214 |
| 17 | 48629 | 199.2024 | 203.0205 | 7.4434 | 26619 | 204.1833 | 9.5900 | 26619 |
| 18 | 123867 | 213.3222 | 217.8203 | 8.5705 | 64057 | 219.1854 | 11.0310 | 64057 |
| 19 | 317955 | 227.4755 | 232.6989 | 9.7500 | 155575 | 234.2795 | 12.5388 | 155575 |
| 20 | 823065 | 241.6596 | 247.6490 | 10.9734 | 381521 | 249.4566 | 14.1024 | 381521 |
| | Parameters | | Lemma 4.1 | | | Lemma 4.2 | | |
| *n* | Count | Avg. | Avg. | Stdev. | Eq. | Avg. | Stdev. | Eq. |
| 10 | 106 | 99.4151 | 113.8491 | 16.2724 | 5 | 164.3774 | 68.8660 | 1 |
| 11 | 235 | 114.4340 | 133.5319 | 21.2783 | 3 | 201.2426 | 91.8521 | 1 |
| 12 | 551 | 128.7477 | 152.3194 | 26.3422 | 5 | 240.4646 | 118.3697 | 1 |
| 13 | 1301 | 142.8932 | 171.5634 | 31.9734 | 2 | 282.1168 | 147.5565 | 1 |
| 14 | 3159 | 156.9839 | 190.7692 | 37.7317 | 10 | 326.6477 | 179.9576 | 1 |
| 15 | 7741 | 171.0559 | 210.3741 | 43.8713 | 2 | 373.8465 | 215.0847 | 1 |
| 16 | 19320 | 185.1062 | 230.0205 | 50.1354 | 9 | 423.8166 | 253.1651 | 1 |
| 17 | 48629 | 199.2024 | 249.9992 | 56.6729 | 6 | 476.5955 | 294.0235 | 1 |
| 18 | 123867 | 213.3222 | 270.1175 | 63.3533 | 16 | 532.1868 | 337.7529 | 1 |
| 19 | 317955 | 227.4755 | 290.4733 | 70.2383 | 2 | 590.5895 | 384.2730 | 1 |
| 20 | 823065 | 241.6596 | 310.9927 | 77.2696 | 27 | 651.8210 | 433.6323 | 1 |

# REFERENCES

1. M. Ali Rostami, H. Martin B?ucker and A. Azadi, Illustrating a Graph Coloring Algorithm Based on the Principle of Inclusion and Exclusion Using GraphTea, LNCS, Springer. 8719 (2014) 514-517.

2. A.R. Ashrafi, T. Došlić and A.Hamzeha, The Zagreb coindices of graph operations, Discrete Appl. Math.158 (2010) 1571-1578.

3. M. Bianchi, A. Cornaro, J. L. Palacios, A. Torriero, New bounds of degree-based topological indices for some classes of c-cyclic graphs. Discrete App. Math. 184 (2015) 62-75.

4. G.B.A. Xavier, E. Suresh and I. Gutman, Counting relations for general Zagreb indices, Kragujevac J. Math. 38 (2014) 95-103.

5. K. C. Das, K. Xu and J. Nam, Zagreb indices of graphs, Front. Math. China 10 (2015) 567-582.

6. N. De, Some bounds of reformulated Zagreb indices, Appl. Math. Sci. 6 (2012) 5005-5012.

7. S. Fajtlowicz S, On conjectures of graffiti II, Congr. Numer. 60 (1987), 189-197.

8. B. Fortula and I. Gutman, A forgotten topological index, J. Math. Chem. 53 (2015) 1184-1190.

9. Gutman, B. Ruščić, N. Trinajstić and C. F. Wilcox, Graph theory and molecular orbitals, XII. Acyclic polyenes, J. Chem. Phys. 62 (1975) 3399-3405.

10. Ilić, M. Ilić and B. Liu, On the upper bounds for the first Zagreb index, Kragujevac Journal of Math. 35 (2011) 173-182.

11. Ilić and B. Zhou, On reformulated Zagreb indices, Discrete App. Math. 160 (2012) 204-209.

12. S. Ji, X. Li and B. Huo, On Reformulated Zagreb Indices with Respect to Acyclic, Unicyclic and Bicyclic Graphs, MATCH Commun. Math. Comput. Chem. 72 (2014) 723{732.

13. X. Li and J. Zheng, A unified approach to the extremal trees for different indices, MATCH Commun. Math. Comput. Chem. 54 (2005) 195-208.

14. T. Mansour and C. Song, The a and (a,b)- Analogs of Zagreb Indices and Coindices of Graphs, Intern. J. Combin. (2012) ID 909285.

15. Miličević, S. Nikolić and N. Trinajstić, On reformulated Zagreb indices, Mol. Divers. 8 (2004) 393-399.

16. E.I. Milovanović, I.Ž. Milovanović, E.Ć. Dolićanin and E. Glogić, A note on the first reformulated Zagreb index, Appl. Math. Comp. 273 (2016) 16-20.

17. G. Su, L. Xiong, L. Xu and B. Ma, On the maximum and minimum first reformulated Zagreb index with connectivity bat most k, FILMAT 25 (2011) 75-83.

18. K. Xu and K.C. Das, Some extremal graphs with respect to inverse degree, Discrete App. Math. (2015).

19. Zhou and N. Trinajstić, On general sum-connectivity index, J. Math. Chem. 47 (2010) 210-218.

20. Zhou and N. Trinajstić, Some properties of the reformulated Zagreb indices, J. Math. Chem. 48 (2010) 714-719.

9

# AN EXTENSIVE ANALYSIS OF THE PARITY OF BROKEN 3-DIAMOND PARTITIONS

**Silviu Radu a, James A. Sellers b**

a Research Institute for Symbolic Computation (RISC), Johannes Kepler University, A-4040 Linz, Austria

b Department of Mathematics, Penn State University, University Park, PA 16802, USA

## ABSTRACT

In 2007, Andrews and Paule introduced the family of functions $\Delta k(n)$ which enumerate the number of broken $k$-diamond partitions for a fixed positive integer $k$. Since then, numerous mathematicians have considered partitions congruences satisfied by $\Delta k(n)$ for small values of $k$. In this work, we provide an extensive analysis of the parity of the function $\Delta 3\ (n)$, including a number of Ramanujan- like congruences modulo 2. This will be accomplished by completely characterizing the values of $\Delta 3\ (8n + r)$ modulo 2 for $r \in \{1, 2, 3, 4, 5, 7\}$ and any value of $n$ 0. In contrast, we conjecture that, for any integers 0 $B$ < $A$, $\Delta 3\ (8(An + B))$ and $\Delta 3\ (8(An + B) + 6)$ is infinitely often even and infinitely often odd. In this sense, we generalize Subbarao's Conjecture for this function $\Delta 3$ . To the best of our knowledge, this is the first generalization of Subbarao's Conjecture in the literature.

## KEYWORDS

Broken *k*-diamonds; Congruences; Modular forms; Partitions

## 1. INTRODUCTION

Broken *k*-diamond partitions were introduced in 2007 by Andrews and Paule [2]. These are constructed in such a way that the generating functions of their counting sequences $(\Delta_k(n))_{n\geqslant 0}$ are closely related to modular forms. Namely,

$$\sum_{n=0}^{\infty} \Delta_k(n)q^n = \prod_{n=1}^{\infty} \frac{(1-q^{2n})(1-q^{(2k+1)n})}{(1-q^n)^3(1-q^{(4k+2)n})}$$
$$= q^{(k+1)/12}\frac{\eta(2\tau)\eta((2k+1)\tau)}{\eta(\tau)^3\eta((4k+2)\tau)}, \quad k \geqslant 1,$$

where we recall the Dedekind eta function

$$\eta(\tau) := q^{\frac{1}{24}} \prod_{n=1}^{\infty} (1-q^n) \quad (q = e^{2\pi i\tau}).$$

In their original work, Andrews and Paule proved that, for all n⩾0n⩾0,

$$\Delta_1(2n+1) \equiv 0 \pmod 3. \tag{1.1}$$

They also conjectured a few other congruences modulo 2 satisfied by certain families of broken *k*-diamond partitions.

Since then, a number of authors have provided proofs of additional congruences satisfied by broken *k*-diamond partitions. Hirschhorn and Sellers [5] provided a new proof of (1.1) above as well as elementary proofs of the following parity results: For all $n\geqslant 0$,

$$\Delta_1(4n+2) \equiv 0 \pmod 2,$$
$$\Delta_1(4n+3) \equiv 0 \pmod 2,$$
$$\Delta_2(10n+2) \equiv 0 \pmod 2, \quad \text{and}$$
$$\Delta_2(10n+6) \equiv 0 \pmod 2$$

The third result in the list above appeared in [2] as a conjecture while the other three did not. Soon after the publication of [5], Chan [3] provided a different proof of the parity results for $\Delta_2\Delta 2$ mentioned above as well as a number of congruences modulo powers of 5. Subsequently, Paule and Radu [7] also proved a number of congruences modulo 5 for broken 2-diamond partitions, and they also shared conjectures related to broken 3-diamond partitions modulo 7 and broken 5-diamond partitions modulo 11. (Two of these conjectures have recently been proven by Xiong [12].)

Our goal in this work is to focus on parity results satisfied by $\Delta_3(n)$. The parity of this function has been studied, at least partially, by Radu and Sellers [10] who proved (among other things) that, for all $n \geqslant 0$,

$$\Delta_3(14n+7) \equiv 0 \pmod 2,$$
$$\Delta_3(14n+9) \equiv 0 \pmod 2, \quad \text{and}$$
$$\Delta_3(14n+13) \equiv 0 \pmod 2. \tag{1.2}$$

We wish to greatly extend results such as those mentioned in (1.2). This will be accomplished by **completely** characterizing the values of $\Delta_3(8n+r)$ modulo 2 for $r \in \{1,2,3,4,5,7\}$ and any value of $n \geqslant 0$ by finding interesting relationships modulo 2 between the generating functions for $\Delta_3(8n+r)$ for these special values of $r$ and classical $q$ -series. We also note here that, while $\Delta_3(8n+r)$ is extremely "well-behaved" modulo 2 for the values $r \in \{1,2,3,4,5,7\}$, and satisfies numerous congruences modulo 2 in arithmetic progressions, we also believe that $\Delta_3$ does **not** satisfy any Ramanujan-like congruences modulo 2 within any subprogression of $8n$ or $8n+68n+6$. In this sense, we generalize Subbarao's Conjecture for this function $\Delta_3$ by calling attention to the two arithmetic progressions $8n$ and $8n+6$. Our hope is that such an analysis will motivate others to complete similar work on other restricted partition functions $f(n)$; namely, to locate a particular value $A$ such that $f(An+r)$ has very nice parity properties for certain values of $r$ while having **no** congruences modulo 2 within the other arithmetic progressions of the form $An+r$. (This seems to be a natural next step in the study of the parity of partition functions given the first author's recent proof of Subbarao's Conjecture [9].)

We note, in passing, that we also prove a number of parity results for $\Delta_3(4n+r)$ and $\Delta_3(2n+r)$ for various values of $r$. We begin with a characterization of the parity of $\Delta_3(2n+1)$ for any $n$.

**Theorem 1.1.**

$$\sum_{n=0}^{\infty} \Delta_3(2n+1)q^n \equiv \prod_{n=1}^{\infty} (1-q^n)(1-q^{7n}) \pmod{2}. \tag{1.3}$$

**Remark 1.2.**

It should be noted that the coefficients of the power series representation of the product on the right-hand side of (1.3) can be completely classified modulo 2. First, we note that

$$q^{1/3}\prod_{n=1}^{\infty} (1-q^n)(1-q^{7n}) \equiv \sum_{m,n\in\mathbb{Z}} q^{\frac{(6m-1)^2+7(6n-1)^2}{24}} \pmod{2}.$$

We then define

$$\sum_{n=0}^{\infty} a(n)q^n := \sum_{m,n\in\mathbb{Z}} q^{(6m-1)^2+7(6n-1)^2} \pmod{2}.$$

Next note that $a(\nu)=0$ unless $\nu = 24k+8$. If $\nu = 24k+8 = 8(3k+1)$ we observe that

$$a(\nu) = \#\{(m,n) \in \mathbb{N}^2 \colon\ m^2 + 7n^2 = 8(3k+1),\ m,n \equiv 1 \pmod 2\}.$$

Moreover, if $7|\nu$, then $a(\nu) = a(\nu/7)$. This is clear because if $m^2 + 7n^2 = 7s$ then $7|m$ which implies that $(7 \cdot (m/7))^2 + 7n^2 = 7s$ which implies that $n^2 + 7(m/7)^2 = s$. Thus every solution to $m^2 + 7n^2 = 7s$ can be transformed into a solution of $m'^2 + 7n'^2 = s$ where $m' = n$ and $n' = m/7$ and vice versa. Next, let $n$ be a positive integer with $7 \nmid n$ and let $\alpha$ be an integer greater than 2. Assume that there exists $x, y \in Z$ with $x, y \equiv 1 \pmod 2$ such that

$$x^2 + 7y^2 = 2^\alpha n.$$

We note that the ring $\mathbb{Z}[\frac{1+\sqrt{-7}}{2}]$is a unique factorization domain. In particular, we have

$$2 = \left(\frac{1+\sqrt{-7}}{2}\right)\left(\frac{1-\sqrt{-7}}{2}\right).$$

Assume that

$$n = p_1^{\alpha_1} \cdots p_s^{\alpha_s} \times (a_1 + \sqrt{-7}b_1)^{\beta_1} \cdots (a_r + \sqrt{-7}b_r)^{\beta_r} \\ \times (a_1 - \sqrt{-7}b_1)^{\beta_1} \cdots (a_r - \sqrt{-7}b_r)^{\beta_r},$$

where $p_j, a_j \pm \sqrt{-7}b_j$ are primes. Set

$$2^\alpha = \left(\frac{1+\sqrt{-7}}{2}\right)^\alpha \left(\frac{1-\sqrt{-7}}{2}\right)^\alpha.$$

Note that

$$(x + \sqrt{-7}y)(x - \sqrt{-7}y) = 2^\alpha n.$$

If $r_j$rj is maximal such that $p_j^{r_j}|(x+\sqrt{-7}y)$, then $p_j^{r_j}|(x+\sqrt{-7}y)$ which implies that $2r_j=\alpha_j$ for j=1,...,sj=1,...,s. It follows that

$$(a_i+\sqrt{-7}b_i)^{j_i}(a_i-\sqrt{-7}b_i)^{\beta_i-j_i}|(x+\sqrt{-7}y),$$

for some $j_i=0,\ldots,\beta_i$ and i=1,...,ri=1,...,r. Furthermore, either

$$\left(\frac{1+\sqrt{-7}}{2}\right)\left(\frac{1-\sqrt{-7}}{2}\right)^{\alpha-1}$$

or

$$\left(\frac{1-\sqrt{-7}}{2}\right)\left(\frac{1+\sqrt{-7}}{2}\right)^{\alpha-1}$$

divides $x+\sqrt{-7}y$.. These are the only possibilities that guarantee that $x$ and $y$ are odd. Consequently, in total we have $2\prod_{j=1}^{r}(1+\beta_j)$ possibilities for $x+\sqrt{-7}y$. If out of this we choose only those with x⩾0 we obtain $\prod_{j=1}^{r}(1+\beta_j)$ possibilities. This implies that

$$a(2^\alpha n)=\prod_{j=1}^{r}(1+\beta_j)$$

where $7\nmid n$ and

$$a(2^\alpha 7^k n)=\prod_{j=1}^{r}(1+\beta_j).$$

Thus,

$$a(2^\alpha 7^k n)\equiv 1 \pmod 2$$

iff $\beta_j$ is even for all $j$ or equivalently if $n$ is a square. Next note that $2^\alpha 7^k t^2\equiv 8 \pmod{24}$ iff $3\nmid t$ and α−3 is even and nonnegative. This implies that

$$\sum_{n=0}^{\infty} a(n)q^n \equiv \sum_{n=0}^{\infty} A(n)q^n \pmod 2,$$

where

$$A(n) := \begin{cases} 1 & \text{if } n = 2^3t^2 \text{ or } n = 2^37t^2,\ 3 \nmid t,\ k \geqslant 0, \\ 0 & \text{otherwise.} \end{cases}$$

Thus

$$q^{1/3}\prod_{n=1}^{\infty}(1-q^n)(1-q^{7n}) \equiv \sum_{n=0}^{\infty} A(n)q^{n/24} = \sum_{k=0}^{\infty} A(24k+8)q^{k+1/3}.$$

Hence,

$$\prod_{n=1}^{\infty}(1-q^n)(1-q^{7n}) \equiv \sum_{k=0}^{\infty} A(24k+8)q^k = \sum_{t\geqslant 0,\, 3\nmid t} q^{\frac{7t^2-1}{3}} + q^{\frac{t^2-1}{3}}.$$

Thanks to the above analysis, we have the following:

**Corollary 1.3.**

*For all* n⩾0; $\Delta_3(2n+1) \equiv 1 \pmod 2$ *if and only if* $3n+1=t^2$ *or* $3n+1=7t^2$.

Notice that the three congruences mentioned in (1.2) follow almost immediately from this characterization given in Corollary 1.3. For example, the above work implies that we need to consider whether 3(7n+3)+13(7n+3)+1 or 21n+1021n+10 can be represented as t²t2 or 7t² for some integer *t* in order to determine the parity of Δ3(14n+7). Note that 21n+1021n+10 is not divisible by 7, so it cannot be written in the form 7t². Moreover, 21n+1021n+10 can never be square because $21n+10 \equiv 3 \pmod 7$and 3 is a quadratic nonresidue modulo 7. In analogous fashion, $\Delta_3(14n+9) \equiv 0 \pmod 2$ because 6 is a quadratic nonresidue modulo 7, and

$\Delta_3(14n+13) \equiv 0 \pmod{2}$ because 5 is a quadratic nonresidue modulo 7.

We now consider parity results satisfied by $\Delta_3(4n+r)$ for various values of *r*.

**Theorem 1.4.**

$$\sum_{n=0}^{\infty} \Delta_3(4n) q^n \equiv \prod_{n=1}^{\infty} \frac{(1-q^n)^5}{(1-q^{7n})} \pmod{2}.$$

**Theorem 1.5.**

$$\sum_{n=0}^{\infty} \Delta_3(4n+2) q^n \equiv q \prod_{n=1}^{\infty} \frac{(1-q^{7n})^5}{(1-q^n)} \pmod{2}.$$

**Theorem 1.6.**

$$\sum_{n=0}^{\infty} \Delta_3(4n+3) q^n \equiv \prod_{n=1}^{\infty} (1-q^{2n})(1-q^{14n}) \pmod{2}.$$

**Remark 1.7.**

A few remarks are in order regarding Theorem 1.6. First, note that the product on the right-hand side of the congruence is an even function of $q$ . This implies that, for all $n \geqslant 0$, $\Delta_3(4(2n+1)+3) \equiv \pmod{2}$ or $\Delta_3(8n+7) \equiv 0 \pmod{2}$ Secondly, note that the right-hand side of Theorem 1.6 is the same as the right-hand side in Theorem 1.1 except with $q$ replaced by $q^2$. Therefore, we can completely characterize the values of $\Delta_3(4n+3)$ modulo 2 via the remarks made regarding Theorem 1.1.

Our last set of theorems provides information about the parity of $\Delta_3(8n+r)$ for a number of values of *r*.

**Theorem 1.8.**

$$\sum_{n=0}^{\infty} \Delta_3(8n+1)q^n \equiv \prod_{n=1}^{\infty}(1-q^{2n}) \pmod{2}.$$

**Remark 1.9.**

As with Theorem 1.6, it is clear that the right-hand side in Theorem 1.8 is an even function of $q$ . Thus, we know that, for all n⩾0, $\Delta_3(16n+9) \equiv 0 \pmod{2}$ immediately. But we actually can say more. Thanks to Euler's Pentagonal Number Theorem [1, Corollary 1.7], we know

$$\prod_{n=1}^{\infty}(1-q^{2n}) = \sum_{m\in\mathbb{Z}}(-1)^m q^{m(3m-1)}.$$

Therefore, we can explicitly state when $\Delta_3$(8n+1) is even or odd; namely, for any n⩾0, $\Delta_3$(8n+1) is odd if and only if n=m(3m−1) for some integer $m$ . This is equivalent to saying $\Delta_3$(8n+1) is odd if and only if 12n+1 is a perfect square. This means we can write down numerous Ramanujan-like congruences modulo 2 within the arithmetic progression 8n+18n+1 with ease.

**Theorem 1.10.**

$$\sum_{n=0}^{\infty} \Delta_3(8n+2)q^n \equiv q\prod_{n=1}^{\infty}(1-q^n)(1-q^{28n}) \pmod{2}.$$

**Theorem 1.11.**

$$\sum_{n=0}^{\infty} \Delta_3(8n+3)q^n \equiv \prod_{n=1}^{\infty}(1-q^n)(1-q^{7n}) \pmod{2}.$$

**Remark 1.12.**

Given Theorem 1.1, we see that Theorem 1.11 clearly implies that, for all n⩾0, $\Delta_3(8n+3) \equiv \Delta_3(2n+1) \pmod{2}$ an attractive "internal" congruence satisfied by $\Delta_3$. We will briefly mention this congruence again in our concluding remarks below.

**Theorem 1.13.**

$$\sum_{n=0}^{\infty} \Delta_3(8n+4)q^n \equiv \prod_{n=1}^{\infty}(1-q^{4n})(1-q^{7n}) \pmod{2}.$$

A remark is in order regarding Theorem 1.10 and Theorem 1.13. We have

$$q^{\frac{1}{24}} \prod_{n=1}^{\infty}(1-q^n) = \sum_{n=-\infty}^{\infty}(-1)^n q^{(6n-1)^2/24} \equiv \sum_{n=-\infty}^{\infty} q^{(6n-1)^2/24} \pmod{2}.$$

Consequently,

$$q^{\frac{11}{24}} \prod_{n=1}^{\infty}(1-q^{4n})(1-q^{7n}) \equiv \sum_{n,m\in\mathbb{Z}} q^{\frac{4(6n-1)^2+7(6m-1)^2}{24}} \pmod{2}$$

and

$$q^{\frac{29}{24}} \prod_{n=1}^{\infty}(1-q^{n})(1-q^{28n}) \equiv \sum_{n,m\in\mathbb{Z}} q^{\frac{(6n-1)^2+28(6m-1)^2}{24}} \pmod{2}.$$

Next we note that

$$n^2+7m^2 \equiv 11 \pmod{24} \quad \Leftrightarrow \quad n=2k,\ k,m\equiv \pm 1 \pmod{6}$$

and

$$n^2+7m^2 \equiv 29 \pmod{24} \quad \Leftrightarrow \quad m=2k,\ n,k\equiv \pm 1 \pmod{6}.$$

For given $x$ with $x \equiv 11 \pmod{24}$ the set of solutions (n,m) such that $4n^2+7m^2=x$ can be partitioned into equivalence classes and two solutions $(n_1,m_1)$ and $(n_2,m_2)$ are equivalent iff $n_1=\pm n_2$ and $m_1=\pm m_2$m1=±m2. In particular each equivalence class has exactly 4 elements and there is only one solution $(n_1,m_1)$ in each class such that $n_1=2k_1$ and $(k_1,m_1) \equiv (-1,-1) \pmod{6}$ This implies in particular that for

$$\sum_{n=0}^{\infty} b(n)q^n := \sum_{n,m\in\mathbb{Z}} q^{n^2+7m^2} \tag{1.4}$$

we have

$$\frac{1}{4}\sum_{n=0}^{\infty} b(24n+11)q^{24n+11} = \sum_{n,m\in\mathbb{Z}} q^{4(6n-1)^2+7(6m-1)^2}.$$

This implies that

$$\frac{1}{4}q^{\frac{11}{24}}\sum_{n=0}^{\infty} b(24n+11)q^{n} = \sum_{n,m\in\mathbb{Z}} q^{\frac{4(6n-1)^2+7(6m-1)^2}{24}}.$$

In a similar fashion we conclude that

$$\frac{1}{4}q^{\frac{29}{24}}\sum_{n=0}^{\infty} b(24n+29)q^{n} = \sum_{n,m\in\mathbb{Z}} q^{\frac{(6n-1)^2+28(6m-1)^2}{24}}.$$

Because of these two relations we observe that in order to understand $\prod_{n=1}^{\infty}(1-q^{4n})\times(1-q^{7n})$ and $\prod_{n=1}^{\infty}(1-q^{n})(1-q^{28n})$ modulo 2 we need to understand b(n)b(n) in (1.4) for $n$ odd. By [4, p. 61, Lemma 3.25] we know that, for m>1m>1 and odd with $7 \nmid m$,

$$|\{x,y\in\mathbb{Z}\colon\ \gcd(x,y)=1,\ x^2+7y^2=m\}| = 2\prod_{p|m}\left(1+\left(\frac{-7}{p}\right)\right).$$

Let $m=m_s^2m_f$ with mfmf squarefree. Then we observe immediately that

$$\left|\{x,y\in\mathbb{Z}\colon\ x^2+7y^2=m\}\right|=2\sum_{d|m_s}\prod_{p|\frac{m}{d^2}}\left(1+\left(\frac{-7}{p}\right)\right).$$

Consequently,

$$b(m)=2\sum_{d|m_s}\prod_{p|\frac{m}{d^2}}\left(1+\left(\frac{-7}{p}\right)\right). \tag{1.5}$$

By using the fact that

$$\left|\{x,y\in\mathbb{Z}\colon\ x^2+7y^2=7^\alpha n\}\right|=\left|\{x,y\in\mathbb{Z}\colon\ x^2+7y^2=n\}\right|$$

one can lift the restriction that 7∤m. From (1.5) we observe that b(m)2 is multiplicative for odd *m*. Because of (1.5), we know for prime p⩾3 that

$$b(p^{2\alpha+1})=2(\alpha+1)\left(1+\left(\frac{-7}{p}\right)\right),$$
$$b(p^{2\alpha})=2\left(\alpha\left(1+\left(\frac{-7}{p}\right)\right)+1\right).$$

This now leads to two corollaries which give a characterization of the values of Δ3(8n+2) and Δ3(8n+2) modulo 2, in terms of this function b(n) just described:

**Corollary 1.14.**

*For all* $all\ n\geqslant 0,\ \Delta_3(8n+2)\equiv\frac{1}{4}b(24n+29)\pmod 2.$

**Corollary 1.15.**

*For* $all\ n\geqslant 0,\ \Delta_3(8n+4)\equiv\frac{1}{4}b(24n+11)\pmod 2.$

**Theorem 1.16.**

$$\sum_{n=0}^{\infty} \Delta_3(8n+5)q^n \equiv \prod_{n=1}^{\infty}(1-q^{14n}) \pmod 2.$$

**Remark 1.17.**

As was discussed after Theorem 1.8, we can employ Euler's Pentagonal Number Theorem here as well to obtain a similar classification result. We can also easily see that, for all n⩾0, $\Delta_3(16n+13) \equiv 0 \pmod 2$ since the right-hand side of Theorem 1.16 is an even function of $q$ . In similar fashion, since the right-hand side is also a function of $q^7$, we can say that, for all n⩾0, $\Delta_3(56n+r) \equiv 0 \pmod 2$ for r∈{13,21,29,37,45,53}

## 2. PROOF OF THE CONGRUENCES

Let

$$f = \sum a(n)q^n := \frac{\eta(6z)\eta(21z)}{\eta^3(3z)\eta(42z)}$$

and

$$\phi := \eta(8z)^{72}.$$

Then

$$a(n) = \Delta_3\left(\frac{n+1}{3}\right).$$

Let

$$g = \sum b(n)q^n.$$

For $\chi$ a character we define

$$g_\chi := \sum \chi(n)b(n)q^n$$

and for $D \in \mathbb{Z},\ \epsilon(n) := \left(\frac{D}{n}\right)$ let

$$g_D := g_\epsilon.$$

Define the UdUd-operator by

$$U_d g := \sum b(dn)q^n.$$

We need that for $F := \sum A(n)q^n,\ G := \sum B(n)q^{nN}$ and $\chi$ a character modulo $N$ we have

$$(FG)_\chi(z) = F_\chi(z)G(z) \tag{2.1}$$

and

$$U_N(FG) = G(z/N)(U_N F)(z). \tag{2.2}$$

One verifies that our congruences are equivalent to the following:

equation(2.3)

$$\begin{aligned}
&\text{Thm. 1.1:} && U_2 f \equiv \eta(3z)\eta(21z) \pmod 2,\\
&\text{Thm. 1.6:} && U_4 f \equiv \eta(6z)\eta(42z) \pmod 2,\\
&\text{Thm. 1.11:} && U_8 f \equiv \eta(3z)\eta(21z) \pmod 2,\\
&\text{Thm. 1.5:} && \frac{1}{2}(f_4 + f_{-4}) \equiv \frac{\eta^5(84z)}{\eta(12z)} \pmod 2,\\
&\text{Thm. 1.4:} && \frac{1}{2}(f_4 - f_{-4}) \equiv \frac{\eta^5(12z)}{\eta(84z)} \pmod 2,\\
&\text{Thm. 1.8:} && \frac{1}{2}([U_2 f]_4 + [U_2 f]_{-4}) \equiv \eta(24z) \pmod 2,\\
&\text{Thm. 1.16:} && \frac{1}{2}([U_2 f]_4 - [U_2 f]_{-4}) \equiv \eta(168z) \pmod 2,\\
&\text{Thm. 1.10:} && \frac{1}{4}(f_4 + f_{-4} - f_8 - f_{-8}) \equiv \eta(24z)\eta(672z) \pmod 2,\\
&\text{Thm. 1.13:} && \frac{1}{4}(f_4 - f_{-4} - f_8 + f_{-8}) \equiv \eta(96z)\eta(168z) \pmod 2.
\end{aligned}$$

Next note that $\phi$ is a series in powers of $q^8$. Let $\phi^{(s)}(z) := \phi(z/s)$. In particular note that $\phi^{(s)}(z)$ is a series in powers of $q^{8/s}$q8/s. Using (2.1) and (2.2) we find

$$\begin{aligned}
U_2(f\phi) &= \phi^{(2)} U_2 f,\\
U_4(f\phi) &\equiv \phi^{(4)} U_4 f,\\
U_8(f\phi) &\equiv \phi^{(8)} U_8 f,\\
\frac{1}{2}((f\phi)_4 + (f\phi)_{-4}) &= \phi \cdot \frac{1}{2}(f_4 + f_{-4}),\\
\frac{1}{2}((f\phi)_4 - (f\phi)_{-4}) &= \phi \cdot \frac{1}{2}(f_4 - f_{-4}),\\
\frac{1}{2}\left(\left[U_2(f\phi)\right]_4 + \left[U_2(f\phi)\right]_{-4}\right) &= \frac{1}{2}\left(\left[\phi^{(2)} U_2 f\right]_4 + \left[\phi^{(2)} U_2 f\right]_{-4}\right)\\
&= \phi^{(2)} \cdot \frac{1}{2}([U_2 f]_4 + [U_2 f]_{-4}),\\
\frac{1}{2}\left(\left[U_2(f\phi)\right]_4 - \left[U_2(f\phi)\right]_{-4}\right) &= \frac{1}{2}\left(\left[\phi^{(2)} U_2 f\right]_4 - \left[\phi^{(2)} U_2 f\right]_{-4}\right)\\
&= \phi^{(2)} \cdot \frac{1}{2}([U_2 f]_4 - [U_2 f]_{-4}),\\
\frac{1}{4}((f\phi)_4 + (f\phi)_{-4} - (f\phi)_8 - (f\phi)_{-8}) &= \phi \cdot \frac{1}{4}(f_4 + f_{-4} - f_8 - f_{-8}),\\
\frac{1}{4}((f\phi)_4 - (f\phi)_{-4} - (f\phi)_8 + (f\phi)_{-8}) &= \phi \cdot \frac{1}{4}(f_4 - f_{-4} - f_8 + f_{-8}).
\end{aligned} \tag{2.4}$$

Recall that $\frac{\eta(\tau)^2}{\eta(2\tau)} \equiv 1 \pmod 2$Then because of (2.4), (2.3) is equivalent to:

$$\begin{aligned}
&\text{Thm. 1.1:} && \left(\frac{\eta(\tau)^2}{\eta(2\tau)}\right)^4 \cdot U_2(f\phi) \equiv \phi^{(2)}\eta(3z)\eta(21z) \pmod 2,\\
&\text{Thm. 1.6:} && \left(\frac{\eta(\tau)^2}{\eta(2\tau)}\right)^4 \cdot U_4(f\phi) \equiv \phi^{(4)}\eta(6z)\eta(42z) \pmod 2,\\
&\text{Thm. 1.11:} && \left(\frac{\eta(\tau)^2}{\eta(2\tau)}\right)^4 \cdot U_8(f\phi) \equiv \phi^{(8)}\eta(3z)\eta(21z) \pmod 2,\\
&\text{Thm. 1.5:} && \left(\frac{\eta(\tau)^2}{\eta(2\tau)}\right)^4 \cdot \frac{1}{2}((f\phi)_4 + (f\phi)_{-4}) \equiv \left(\frac{\eta(\tau)^2}{\eta(2\tau)}\right)^{-2} \cdot \phi \cdot \frac{\eta^5(84z)}{\eta(12z)} \pmod 2,\\
&\text{Thm. 1.4:} && \left(\frac{\eta(\tau)^2}{\eta(2\tau)}\right)^4 \cdot \frac{1}{2}((f\phi)_4 - (f\phi)_{-4}) \equiv \left(\frac{\eta(\tau)^2}{\eta(2\tau)}\right)^{-2} \cdot \phi \cdot \frac{\eta^5(12z)}{\eta(84z)}\\
& && \pmod 2,\\
&\text{Thm. 1.8:} && \left(\frac{\eta(\tau)^2}{\eta(2\tau)}\right)^4 \cdot \frac{1}{2}\left(\left[U_2(f\phi)\right]_4 + \left[U_2(f\phi)\right]_{-4}\right) \equiv \phi^{(2)} \cdot \eta(24z)\\
& && \equiv \phi^{(2)} \cdot \eta(12z)^2 \pmod 2,\\
&\text{Thm. 1.16:} && \left(\frac{\eta(\tau)^2}{\eta(2\tau)}\right)^4 \cdot \frac{1}{2}\left(\left[U_2(f\phi)\right]_4 - \left[U_2(f\phi)\right]_{-4}\right) \equiv \phi^{(2)} \cdot \eta(168z)\\
& && \equiv \phi^{(2)} \cdot \eta(84z)^2 \pmod 2,\\
&\text{Thm. 1.10:} && \left(\frac{\eta(\tau)^2}{\eta(2\tau)}\right)^4 \cdot \frac{1}{4}((f\phi)_4 + (f\phi)_{-4} - (f\phi)_8 - (f\phi)_{-8})\\
& && \equiv \phi \cdot \eta(24z)\eta(672z) \pmod 2,\\
&\text{Thm. 1.13:} && \left(\frac{\eta(\tau)^2}{\eta(2\tau)}\right)^4 \cdot \frac{1}{4}((f\phi)_4 - (f\phi)_{-4} - (f\phi)_8 + (f\phi)_{-8})\\
& && \equiv \phi \cdot \eta(96z)\eta(168z) \pmod 2.
\end{aligned} \tag{2.5}$$

Denote by $M_k(N,\chi)$ the set of weak modular forms of weight $k$ and character $\chi$ for the group $\Gamma_0(N)$. By [6, Thm. 1.64] we have that $f\phi \in M_{35}(504, (\frac{-1}{d}))$ and $(\frac{\eta(\tau)^2}{\eta(2\tau)})^4 \in M_2(4, \mathrm{id})$ Furthermore, by [6, Prop. 2.8] we have that if $g \in M_k(N,\chi)$ and $(\frac{D}{n})$ is a character modulo $m$ , then $g_D \in M_k(Nm^2,\chi)$. By [6, Prop. 2.22], if $g \in M_k(N,\chi)$) and $d|N$, then $U_d f \in M_k(N,\chi)$ This implies that the left-hand side of the relations in (2.5) in the first three lines are in $M_{37}(504, (\frac{-1}{d}))$ in the next four lines they are in $M_{37}(504 \cdot 4^2, (\frac{-1}{d}))$ and in the last two lines they are in

$M_{37}(504 \cdot 8^2, (\frac{-1}{d}))$ One can check the same holds for the functions on the right-hand side using [6, Thm. 1.64]. Using a generalization of Sturm's theorem [11], namely [6, Thm. 2.58], we find that the first three identities hold if they hold for the first $\frac{35}{12} \times 1152 \sim 3360$ coefficients in their $q$ -expansion. Similarly, the next four identities hold if they hold for $\frac{35}{12} \times 18432 \sim 53760$coefficients in their $q$-expansions. Finally for the last two identities on needs to check about 215 040 coefficients modulo 2.

**Remark 2.1.**

An alternative method to prove these identities in their original form is using the approach from [8] which leads to a less elegant proof but more direct on the problem. We calculated using this method that we do not need to compute more than 1056 coefficients for any of the identities.

# 3. CLOSING COMMENTS

We close this note by sharing a conjectured infinite family of "internal" congruences satisfied by $\Delta_3(n)\Delta3(n)$ modulo powers of 2:

**Conjecture.**

Let

$$\lambda_\alpha = \begin{cases} \frac{2^{\alpha+1}+1}{3} & \textit{if } \alpha \textit{ is even,} \\ \frac{2^\alpha+1}{3} & \textit{if } \alpha \textit{ is odd.} \end{cases}$$

Then, for all $\alpha \geqslant 1$and $n \geqslant 0$,

$$\Delta_3(\lambda_\alpha)\Delta_3(2^{\alpha+2}n + \lambda_{\alpha+2}) \equiv \Delta_3(\lambda_{\alpha+2})\Delta_3(2^\alpha n + \lambda_\alpha) \pmod{2^\alpha}$$

and

$$\Delta_3(\lambda_\alpha) \equiv 1 \pmod 2.$$

The case α=1 of this conjecture was proven above; namely, in Remark *1.12*, we noted that

$$\Delta_3(8n+3) \equiv \Delta_3(2n+1) \pmod 2$$

for all n⩾0n.

## ACKNOWLEDGMENTS

The authors thank Heinrich Rolletschek for helpful discussions related to algebraic number theory. The authors also gratefully acknowledge the referee for insightful comments related to the proof techniques found in this paper.

## REFERENCES

1. G.E. Andrews, The Theory of Partitions, Addison–Wesley, 1976.
2. G.E. Andrews, P. Paule, MacMahon's partition analysis XI: Broken diamonds and modular forms, Acta Arith. 126 (2007) 281–294.
3. S.H. Chan, Some congruences for Andrews–Paule's broken 2-diamond partitions, Discrete Math. 308 (2008) 5735–5741.
4. D.A. Cox, Primes of the Form x2 + ny2: Fermat, Class Field Theory, and Complex Multiplication, John Wiley and Sons, 1989.
5. M.D. Hirschhorn, J.A. Sellers, On recent congruence results of Andrews and Paule, Bull. Aust. Math. Soc. 75 (2007) 121–126.

6. K. Ono, The Web of Modularity: Arithmetic of the Coefficients of Modular Forms and q-Series, CBMS Regional Conference Series in Mathematics, vol. 102, Washington, DC, 2004, published for the Conference Board of the Mathematical Sciences.

7. P. Paule, S. Radu, Infinite families of strange partition congruences for broken 2-diamonds, Ramanujan J. 23 (1–3) (2010) 409–416.

8. S. Radu, An algorithmic approach to Ramanujan's congruences, Ramanujan J. 20 (2) (2009) 215–251.

9. S. Radu, A proof of Subbarao's conjecture, J. Reine Angew. Math. 672 (2012) 161–175.

10. S. Radu, J.A. Sellers, Parity results for broken k-diamond partitions and (2k + 1)-cores, Acta Arith. 146 (2011) 43–52.

11. J. Sturm, On the Congruence of Modular Forms, Lecture Notes in Math., vol. 1240, Springer, Berlin, Heidelberg, 1987, pp. 275–280.

12. X. Xiong, Two congruences involving Andrews–Paule's broken 3-diamond partitions and 5-diamond partitions, Proc. Japan Acad. Ser. A Math. Sci. 87 (5) (2011) 65–68.

10

# MULTIPLE TWISTED $q$-EULER NUMBERS AND POLYNOMIALS ASSOCIATED WITH $p$-ADIC $q$-INTEGRALS


## ABSTRACT

By using $p$-adic $q$-integrals on $\mathbb{Z}_p$, we define multiple twisted $q$-Euler numbers and polynomials. We also find Witt's type formula for multiple twisted $q$-Euler numbers and discuss some characterizations of multiple twisted $q$-Euler Zeta functions. In particular, we construct multiple twisted Barnes' type $q$-Euler polynomials and multiple twisted Barnes' type $q$-Euler Zeta functions. Finally, we define multiple twisted Dirichlet's type $q$-Euler numbers and polynomials, and give Witt's type formula for them.


## 1. INTRODUCTION

Let $p$ be a fixed odd prime number. Throughout this paper, $Z_p$, $Q_p$, and $C_p$ are, respectively, the ring of $p$-adic rational integers, the field of $p$-adic rational numbers, and the $p$-adic completion of the algebraic closure of $Q_p$. The $p$-adic absolute value in $C_p$ is normalized so that $|p|_p = 1/p$. When one talks about $q$-extension, $q$ is variously considered as an indeterminate, a complex number, $q \in C$ or a $p$-adic number $q \in C_p$. If $q \in C$, one normally assumes that $|q| < 1$. If $q \in C_p$, one normally assumes that $|1-q|_p < p^{-1/(p-1)}$ so that $q^x = \exp(x \log q)$ for each $x \in Z_p$. We use the notations

$$[x]_q = \frac{1-q^x}{1-q}, \qquad [x]_{-q} = \frac{1-(-q)^x}{1+q} \tag{1.1}$$

(cf. [1, 2, 3, 4, 5, 6, 7, 8, 9, 10, 11, 12, 13, 14]), for all $x \in Z_p$. For a fixed odd positive integer $d$ with $(p,d)=1$, set

$$\begin{aligned} X &= X_d = \varprojlim_{n} Z/dp^nZ, \qquad X_1 = Z_p, \\ X^* &= \bigcup_{\substack{0<a<dp \\ (a,p)=1}} (a+dpZ_p), \\ a+dp^nZ_p &= \{x \in X \mid x \equiv a \pmod{dp^n}\}, \end{aligned} \tag{1.2}$$

where $a \in Z$ lies in $0 \le a < dp^n$. For any $n \in N$,

$$\mu_q(a+dp^nZ_p) = \frac{q^a}{[dp^n]_q} \tag{1.3}$$

is known to be a distribution on $X$ (cf. [1, 2, 3, 4, 5, 6, 7, 8, 9, 10, 11, 12, 13, 14, 15, 16, 17, 18, 19, 20, 21, 22, 23, 24, 25, 26, 27, 28]).

We say that $f$ is uniformly differentiable function at a point $a \in Z_p$ and denote this property by $f \in UD(Z_p)$ if the difference quotients

$$F_f(x,y) = \frac{f(x)-f(y)}{x-y} \tag{1.4}$$

have a limit $l = f'(a)$ as $(x,y) \to (a,a)$ (cf. [25]).

The $p$-adic $q$-integral of a function $f \in UD(Z_p)$ was defined as

$$I_q(f) = \int_{Z_p} f(x)d\mu_q(x) = \lim_{n\to\infty} \frac{1}{[p^n]_q} \sum_{x=0}^{p^n-1} f(x)q^x, \tag{1.5}$$

$$I_{-q}(f) = \int_{Z_p} f(x)d\mu_{-q}(x) = \lim_{n\to\infty} \frac{1}{[p^n]_q} \sum_{x=0}^{p^n-1} f(x)(-q)^x, \tag{1.6}$$

(cf. [4, 24, 25, 28]), from (1.6), we derive

$$qI_{-q}(f_1)+I_{-q}(f)=[2]_q f(0), \tag{1.7}$$

where $f_1(x)=f(x+1)$. If we take $f(x)=e^{tx}$, then we have $f_1(x)=e^{t(x+1)}=e^{tx}e^t$. From (1.7), we obtain that

$$I_{-q}(e^{tx})=\frac{[2]_q}{qe^t+1}. \tag{1.8}$$

In Section 2, we define the multiple twisted $q$-Euler numbers and polynomials on $Z_p$ and find Witt's type formula for multiple twisted $q$-Euler numbers. We also have sums of consecutive multiple twisted $q$-Euler numbers. In Section 3, we consider multiple twisted $q$-Euler Zeta functions which interpolate new multiple twisted $q$-Euler polynomials at negative integers and investigate some characterizations of them. In Section 4, we construct the multiple twisted Barnes' type $q$-Euler polynomials and multiple twisted Barnes' type $q$-Euler Zeta functions which interpolate new multiple twisted Barnes' type $q$-Euler polynomials at negative integers. In Section 5, we define multiple twisted Dirichlet's type $q$-Euler numbers and polynomials and give Witt's type formula for them.

## 2. MULTIPLE TWISTED $q$-EULER NUMBERS AND POLYNOMIALS

In this section, we assume that $q\in C_p$ with $|1-q|_p<1$. For $n\in N$, by the definition of $p$-adic $q$-integral on $Z_p$, we have

$$q^n I_{-q}(f_n)+(-1)^{n-1}I_{-q}(f)=[2]_q\sum_{x=0}^{n-1}(-1)^{n-1-x}q^x f(x), \tag{2.1}$$

where $f_n(x)=f(x+n)$. If $n$ is odd positive integer, we have

$$q^n I_{-q}(f_n)+I_{-q}(f)=[2]_q\sum_{x=0}^{n-1}(-1)^{n-1-x}q^x f(x). \tag{2.2}$$

Let $T_p = \cup_{n\geq 1} C_{p^n} = \lim_{n\to\infty} C_{p^n} = C_{p^\infty}$ be the locally constant space, where $C_{p^n} = \{w \mid w^{p^n} = 1\}$ is the cyclic group of order $p^n$. For $w \in T_p$, we denote the locally constant function by

$$\phi_w : Z_p \to C_p,\ x \to w^x, \tag{2.3}$$

(cf. [5, 7, 8, 9, 10, 11, 12, 13, 14, 16, 18]). If we take $f(x) = \phi_w(x)e^{tx}$, then we have

$$\int_{Z_p} e^{tx}\phi_w(x)d\mu_{-q}(x) = \frac{[2]_q}{qwe^t+1}. \tag{2.4}$$

Now we define the twisted $q$-Euler numbers $E^q_{n,w}$ as follows:

$$F_w(t) = \frac{[2]_q}{qwe^t+1} = \sum_{n=o}^{\infty} E^q_{n,w}\frac{t^n}{n!}. \tag{2.5}$$

We note that by substituting $w = 1$, $\lim_{q\to 1} E^q_{n,1} = E_n$ are the familiar Euler numbers. Over five decades ago, Carlitz defined $q$-extension of Euler numbers (cf. [15]). From (2.4) and (2.5), we note that Witt's type formula for a twisted $q$-Euler number is given by

$$\int_{Z_p} x^n w^x d\mu_{-q}(x) = E^q_{n,w}. \tag{2.6}$$

for each $w \in T_p$ and $n \in N$.

Twisted $q$-Euler polynomials $E^q_{n,w}(x)$ are defined by means of the generating function

$$F^q_w(t,x) = \frac{[2]_q}{qwe^t+1}e^{xt} = \sum_{n=0}^{\infty} E^q_{n,w}(x)\frac{t^n}{n!}, \tag{2.7}$$

where $E^q_{n,w}(0) = E^q_{n,w}$. By using the $h$ th iterative fermionic $p$-adic $q$-integral on $Z_p$, we define multiple twisted $q$-Euler number as follows:

$$\underbrace{\int_{Z_p}\cdots\int_{Z_p}}_{h\text{-times}} w^{x_1+\cdots+x_h} e^{(x_1+x_2+\cdots+x_h)t} d\mu_{-q}(x_1)\cdots d\mu_{-q}(x_h) = \left(\frac{[2]_q}{qwe^t+1}\right)^h = \sum_{n=0}^{\infty} E_{n,w}^{(h,q)} \frac{t^n}{n!}. \tag{2.8}$$

Thus we give Witt's type formula for multiple twisted $q$-Euler numbers as follows.

Theorem 2.1.

For each $w \in T_p$ and $h, n \in N$,

$$\underbrace{\int_{Z_p}\cdots\int_{Z_p}}_{h\text{-times}} w^{x_1+\cdots+x_h} (x_1+\cdots+x_h)^n d\mu_{-q}(x_1)\cdots d\mu_{-q}(x_h) = E_{n,w}^{(h,q)}, \tag{2.9}$$

where

$$(x_1+\cdots+x_h)^n = \sum_{\substack{l_1+\cdots+l_h=n \\ l_1,\ldots,l_h \geq 0}} \frac{n!}{l_1!\cdots l_h!} x_1^{l_1}\cdots x_h^{l_h}. \tag{2.10}$$

From (2.8) and (2.9), we obtain the following theorem.

Theorem 2.2.

For $w \in T_p$ and $h, k \in N$,

$$E_{k,w}^{(h,q)} = \sum_{\substack{l_1+\cdots+l_h=k \\ l_1,\ldots,l_h \geq 0}} \frac{k!}{l_1!\cdots l_h!} E_{l_1,w}^{q} \cdots E_{l_h,w}^{q}. \tag{2.11}$$

From these formulas, we consider multivariate fermionic $p$-adic $q$-integral on $Z_p$ as follows:

$$\underbrace{\int_{Z_p}\cdots\int_{Z_p}}_{h\text{-times}} w^{x_1+\cdots+x_h} e^{(x_1+\cdots+x_h+x)t} d\mu_{-q}(x_1)\cdots d\mu_{-q}(x_h)$$

$$=\left(\frac{[2]_q}{qwe^t+1}\right)\cdots\left(\frac{[2]_q}{qwe^t+1}\right)e^{xt}=\left(\frac{[2]_q}{qwe^t+1}\right)^h e^{xt}. \qquad (2.12)$$

Then we can define the multiple twisted $q$-Euler polynomials $E_{n,w}^{(h,q)}(x)$ as follows:

$$F_w^{(h,q)}(t,x)=\left(\frac{[2]_q}{qwe^t+1}\right)^h e^{xt}=\sum_{n=0}^{\infty} E_{n,w}^{(h,q)}(x)\frac{t^n}{n!}. \qquad (2.13)$$

From (2.12) and (2.13), we note that

$$\sum_{n=0}^{\infty}\underbrace{\int_{Z_p}\cdots\int_{Z_p}}_{h\text{-times}} w^{x_1+\cdots+x_h}(x_1+\cdots+x_h+x)^n d\mu_{-q}(x_1)\cdots d\mu_{-q}(x_h)\frac{t^n}{n!}=\sum_{n=0}^{\infty} E_{n,w}^{(h,q)}(x)\frac{t^n}{n!}.$$

$$(2.14)$$

Then by the $k$th differentiation on both sides of (2.14), we obtain the following.

Theorem 2.3.

For each $w \in T_p$ and $k,h \in N$,

$$\underbrace{\int_{Z_p}\cdots\int_{Z_p}}_{h\text{-times}} w^{x_1+\cdots+x_h}(x_1+\cdots+x_h+x)^k d\mu_{-q}(x_1)\cdots d\mu_{-q}(x_h)=E_{k,w}^{(h,q)}(x). \qquad (2.15)$$

Note that

$$(x_1+\cdots+x_h+x)^n=\sum_{\substack{l_1+\cdots+l_h=n\\ l_1,\ldots,l_h\geq 0}}\frac{n!}{l_1!\cdots l_h!}x_1^{l_1}\cdot x_2^{l_2}\cdots(x_h+x)^{l_h}. \qquad (2.16)$$

Then we see that

$$\underbrace{\int_{Z_p}\cdots\int_{Z_p}}_{h\text{-times}} w^{x_1+\cdots+x_h}(x_1+\cdots+x_h+x)^k d\mu_{-q}(x_1)\cdots d\mu_{-q}(x_h)$$

$$= \sum_{\substack{l_1+\cdots+l_h=k \\ l_1,\ldots,l_h\geq 0}} \frac{k!}{l_1!\cdots l_h!}\int_{Z_p} w^{x_1}x_1^{l_1}d\mu_{-q}(x_1)$$

$$\cdots\int_{Z_p} w^{x_{h-1}}x_{h-1}^{l_{h-1}}d\mu_{-q}(x_{h-1})\int_{Z_p}(x+x_h)^{l_h}d\mu_{-q}(x_h)$$

$$= \sum_{\substack{l_1+\cdots+l_h=k \\ l_1,\ldots,l_h\geq 0}} \frac{k!}{l_1!\cdots l_h!}E_{l_1,w}^{q}\cdots E_{l_{h-1},w}^{q}E_{l_h,w}^{q}(x). \tag{2.17}$$

From (2.15) and (2.17), we obtain the sums of powers of consecutive $q$-Euler numbers as follows.

Theorem 2.4.

For each $w \in T_p$ and $k, h \in N$,

$$E_{k,w}^{(h,q)}(x) = \sum_{\substack{l_1+\cdots+l_h=k \\ l_1,\ldots,l_h\geq 0}} \frac{k!}{l_1!\cdots l_h!}E_{l_1,w}^{q}\cdots E_{l_{h-1},w}^{q}\cdot E_{l_h,w}^{q}(x). \tag{2.18}$$

# 3. MULTIPLE TWISTED $q$-EULER ZETA FUNCTIONS

For $q \in C$ with $|q|<1$ and $w \in T_p$, the multiple twisted $q$-Euler numbers can be considered as follows:

$$F_w^h(t) = \left(\frac{[2]_q}{qwe^t+1}\right)^h = \sum_{n=0}^{\infty} E_{n,w}^{(h,q)}\frac{t^n}{n!}, \quad |t+\log(qw)|<\pi. \tag{3.1}$$

From (3.1), we note that

$$\sum_{n=0}^{\infty} E_{n,w}^{(h,q)} \frac{t^n}{n!} = F_w^h(t) = \left(\frac{[2]_q}{qwe^t+1}\right)^h = [2]_q^h \left(\frac{[2]_q}{qwe^t+1}\right) \cdots \left(\frac{[2]_q}{qwe^t+1}\right)$$
$$= [2]_q^h \sum_{n_1=0}^{\infty} (-1)^{n_1} q^{n_1} w^{n_1} e^{n_1 t} \cdots \sum_{n_h=0}^{\infty} (-1)^{n_h} q^{n_h} w^{n_h} e^{n_h t}$$
$$= [2]_q^h \sum_{n_1,\ldots,n_h=0} (-1)^{n_1+\cdots+n_h} q^{n_1+\cdots+n_h} w^{n_1+\cdots+n_h} e^{(n_1+\cdots+n_h)t}. \quad (3.2)$$

By the $k$th differentiation on both sides of (3.2) at $t = 0$, we obtain that

$$E_{k,w}^{(h,q)} = [2]_q^h \sum_{\substack{n_1+\cdots+n_h \neq 0 \\ n_1,\ldots,n_h \geq 0}} (-1)^{n_1+\cdots+n_h} q^{n_1+\cdots+n_h} w^{n_1+\cdots+n_h} (n_1+\cdots+n_h)^k. \quad (3.3)$$

From (3.3), we derive multiple twisted $q$-Euler Zeta function as follows:

$$\zeta_w^{(h,q)}(s) = [2]_q^h \sum_{\substack{n_1+\cdots+n_h \neq 0 \\ n_1,\ldots,n_h \geq 0}} \frac{(-1)^{n_1+\cdots+n_h} q^{n_1+\cdots+n_h} w^{n_1+\cdots+n_h}}{(n_1+\cdots+n_h)^s} \quad (3.4)$$

for all $s \in C$. We also obtain the following theorem in which multiple twisted $q$-Euler Zeta functions interpolate multiple twisted $q$-Euler polynomials.

Theorem 3.1.

For $w \in T_p$ and $k, h \in N$,

$$\zeta_w^{(h,q)}(-k) = E_{k,w}^{(h,q)}. \quad (35)$$

## 4. MULTIPLE TWISTED BARNES' TYPE $q$-EULER POLYNOMIALS

In this section, we consider the generating function of multiple twisted $q$-Euler polynomials:

$$F_w^h(t,x) = \left(\frac{[2]_q}{qwe^t+1}\right)^h e^{xt} = \sum_{n=0}^{\infty} E_{n,w}^{(h,q)}(x)\frac{t^n}{n!},$$
$$|t+\log(qw)| < \pi, \qquad \mathrm{Re}(x) > 0. \tag{4.1}$$

We note that

$$\sum_{n=0}^{\infty} E_{n,w}^{(h,q)}(x)\frac{t^n}{n!} = F_w^h(t,x) = [2]_q^h \sum_{n_1,\dots,n_h=0}^{\infty} (-1)^{n_1+\cdots+n_h} q^{n_1+\cdots+n_h} w^{n_1+\cdots+n_h} e^{(n_1+\cdots+n_h+x)t}. \tag{4.2}$$

By the $k$th differentiation on both sides of (4.2) at $t=0$, we obtain that

$$E_{k,w}^{(h,q)}(x) = [2]_q^h \sum_{n_1,\dots,n_h=0}^{\infty} (-1)^{n_1+\cdots+n_h} q^{n_1+\cdots+n_h} w^{n_1+\cdots+n_h} (n_1+\cdots+n_h+x)^k. \tag{4.3}$$

Thus we can consider multiple twisted Hurwitz's type $q$-Euler Zeta function as follows:

$$\zeta_w^{(h,q)}(s,x) = [2]_q^h \sum_{\substack{n_1+\cdots+n_h \neq 0 \\ n_1,\dots,n_h \geq 0}} \frac{(-1)^{n_1+\cdots+n_h} q^{n_1+\cdots+n_h} w^{n_1+\cdots+n_h}}{(n_1+\cdots+n_h+x)^s} \tag{4.4}$$

for all $s \in \mathbb{C}$ and $\mathrm{Re}(x) > 0$. We note that $\zeta_w^{(h,q)}(s,x)$ is analyticc function in the whole complex $s$-plane and $\zeta_w^{(h,q)}(s,0) = \zeta_w^{(h,q)}(s)$. We also remark that if $w=1$ and $h=1$, then $\zeta_1^{(1,q)}(s,x) = \zeta^q(s,x)$ is Hurwitz's type $q$-Euler Zeta function (see [7, 27]). The following theorem means that multiple twisted $q$-Euler Zeta functions interpolate multiple twisted $q$-Euler polynomials at negative integers.

Theorem 4.1.

For $w \in T_p$, $k,h \in N$, $s \in C$, and $\mathrm{Re}(x) > 0$,

$$\zeta_w^{(h,q)}(-k,x) = E_{k,w}^{(h,q)}(x). \tag{4.5}$$

Let us consider

$$\begin{aligned} F_w^h(a_1,\dots,a_h \mid t,x) &= \left(\frac{[2]_q}{qwe^{a_1t}+1}\right)\cdots\left(\frac{[2]_q}{qwe^{a_ht}+1}\right)e^{xt} \\ &= [2]_q^h \sum_{n_1,\dots,n_h=0}^{\infty} (-1)^{n_1+\dots+n_h} q^{n_1+\dots+n_h} w^{n_1+\dots+n_h} e^{(a_1n_1+\dots+a_hn_h+x)t} \\ &= \sum_{n=0}^{\infty} E_{n,w}^{(h,q)}(a_1,\dots,a_h \mid x)\frac{t^n}{n!}, \end{aligned} \tag{4.6}$$

where $a_1,\dots,a_h \in C$ and $\max_{1\le i\le k}\{|\log(q+a_it)|\} < n$. Then $E_{n,w}^{(h,q)}(a_1,\dots,a_h \mid x)$ will be called multiple twisted Barnes' type $q$-Euler polynomials. We note that

$$E_{n,w}^{(h,q)}(1,1,\dots,1 \mid x) = E_{n,w}^{(h,q)}(x). \tag{4.7}$$

By the $k$th differentiation of both sides of (4.6), we obtain the following theorem.

Theorem 4.2.

For each $w \in T_p$, $a_1,\dots,a_h \in C$, $k,h \in N$, and $\mathrm{Re}(x) > 0$,

$$E_{k,w}^{(h,q)}(a_1,\dots,a_h \mid x) = [2]_q^h \sum_{\substack{n_1+\dots+n_h \neq 0 \\ n_1,\dots,n_h \ge 0}} (-1)^{n_1+\dots+n_h} q^{n_1+\dots+n_h} w^{n_1+\dots+n_h} (a_1n_1+\dots+a_hn_h+x)^k, \tag{4.8}$$

where

$$(a_1n_1+\dots+a_hn_h+x)^k = \sum_{\substack{l_1+\dots+l_h=k \\ l_1,\dots,l_h \ge 0}} \frac{k!}{l_1!\cdots l_h!} a_1^{l_1}\cdots a_{h-1}^{l_{h-1}} n_1^{l_1}\cdots n_{h-1}^{l_{h-1}} (a_hn_h+x)^{l_h}. \tag{4.9}$$

From (4.8), we consider multiple twisted Barnes' type $q$-Euler Zeta function defined as follows: for each $w \in T_p$, $a_1,\dots,a_h \in C$, $k,h \in N$, and $\mathrm{Re}(x) > 0$,

$$\zeta_{k,w}^{(h,q)}(a_1,\ldots,a_h \mid s,x)=[2]_q^h \sum_{\substack{n_1+\cdots+n_h\neq 0\\ n_1,\ldots,n_h\geq 0}} \frac{(-1)^{n_1+\cdots+n_h}q^{n_1+\cdots+n_h}w^{n_1+\cdots+n_h}}{(a_1n_1+\cdots+a_hn_h+x)^s}. \tag{4.10}$$

We note that $\zeta_{k,w}^{(h,q)}(a_1,\ldots,a_h \mid s,x)$ is analytic function in the whole complex $s$-plane. We also see that multiple twisted Barnes' type $q$-Euler Zeta functions interpolate multiple twisted Barnes' type $q$-Euler polynomials at negative integers as follows.

Theorem 4.3.

For each $w\in T_p$, $a_1,\ldots,a_h\in C$, $k,h\in N$, and $\mathrm{Re}(x)>0$,

$$\zeta_{k,w}^{(h,q)}(a_1,\ldots,a_h \mid -k,x)=E_{k,w}^{(h,q)}(a_1,\ldots,a_h \mid x). \tag{4.11}$$

# 5. MULTIPLE TWISTED DIRICHLET'S TYPE $q$-EULER NUMBERS AND POLYNOMIALS

Let $\chi$ be a Dirichlet's character with conductor $d(=\text{odd})\in N$ and $w\in T_p$. If we take $f(x)=\chi(x)\phi_w(x)e^{tx}$, then we have $f_d(x)=f(x+d)=\chi(x)w^de^{td}w^xe^{tx}$. From (2.2), we derive

$$\int_X \chi(x)w^xe^{tx}d\mu_{-q}(x)=\frac{[2]_q\sum_{i=0}^{d-1}(-1)^{d-1-i}q^i\chi(i)w^ie^{ti}}{q^dw^de^{td}+1}. \tag{5.1}$$

In view of (5.1), we can define twisted Dirichlet's type $q$-Euler numbers as follows:

$$\begin{aligned} F_{w,\chi}^q(t) &= \frac{[2]_q\sum_{i=0}^{d-1}(-1)^{d-1-i}q^i\chi(i)w^ie^{ti}}{q^dw^de^{td}+1}\\ &=\sum_{n=0}^{\infty}E_{n,\chi,w}^q\frac{t^n}{n!},\quad |t+\log(qw)|<\frac{\pi}{d}, \end{aligned} \tag{5.2}$$

(cf. [17, 19, 21, 22]). From (5.1) and (5.2), we can give Witt's type formula for twisted Dirichlet's type $q$-Euler numbers as follows.

Theorem 5.1.

Let $\chi$ be a Dirichlet's character with conductor $d(=\text{odd}) \in N$. For each $w \in T_p$, $n \in N \cup \{0\}$, we have

$$\int_X \chi(x) w^x e^{tx} d\mu_{-q}(x) = E^q_{n,\chi,w}. \tag{5.3}$$

We note that if $w = 1$, then $E^q_{n,\chi,1} = E^q_{n,\chi}$ is the generalized $q$-Euler numbers attached to $\chi$ (see [18, 26]). From (5.2), we also see that

$$\begin{aligned} F^q_{w,\chi}(t) &= [2]_q \sum_{i=0}^{d-1} (-1)^{d-1-i} q^i \chi(i) w^i e^{ti} \sum_{l=0}^{\infty} q^{ld} w^{ld} e^{ldt} (-1)^l \\ &= [2]_q \sum_{n=0}^{\infty} (-1)^n q^n w^n \chi(n) e^{nt}. \end{aligned} \tag{5.4}$$

By (5.2) and (5.4), we obtain that

$$E^q_{k,\chi,w} = \frac{d^k}{dt^k} F^q_{w,\chi}(t)\ |_{t=0} = [2]_q \sum_{n=0}^{\infty} (-1)^n q^n w^n \chi(n) n^k. \tag{5.5}$$

From (5.5), we can define the $l^q_{w,\chi}$-function as follows:

$$l^q_{\chi,w}(s) = [2]_q \sum_{n=0}^{\infty} \frac{(-1)^n q^n w^n \chi(n)}{n^s} \tag{5.6}$$

for all $s \in C$. We note that $l^q_{\chi,w}(s)$ is analytic function in the whole complex $s$-plane. From (5.5) and (5.6), we can derive the following result.

Theorem 5.2.

Let $\chi$ be a Dirichlet's character with conductor $d(=\text{odd}) \in N$. For each $w \in T_p$, $n \in N \cup \{0\}$, we have

$$l^q_{w,\chi}(-n) = E^q_{n,\chi,w}. \tag{5.7}$$

Now, in view of (5.1), we can define multiple twisted Dirichlet's type $q$-Euler numbers by means of the generating function as follows:

$$F_{w,\chi}^{(h,q)}(t) = \left(\frac{[2]_q \sum_{i=0}^{d-1}(-1)^{d-1-i} q^i \chi(i) w^i e^{ti}}{q^d w^d e^{td}+1}\right)^h$$
$$= \left(\int_X \chi(x) w^x e^{tx} d\mu_{-q}(x)\right)^h = \sum_{n=0}^{\infty} E_{n,\chi,w}^{(h,q)} \frac{t^n}{n!}, \tag{5.8}$$

where $|t+\log(qw)| < \pi/d$. We note that if $w=1$, then $E_{n,\chi,1}^{q}$ is a multiple generalized $q$-Euler number (see [22]).

By using the same method used in (2.8) and (2.9),

$$\sum_{n=0}^{\infty} \underbrace{\int_X \cdots \int_X}_{h\text{-times}} \chi(x_1+\cdots+x_h) w^{x_1+\cdots+x_h}(x_1+\cdots+x_h)^n d\mu_{-q}(x_1)\cdots d\mu_{-q}(x_h)\frac{t^n}{n!} = \sum_{n=0}^{\infty} E_{n,w}^{(h,q)} \frac{t^n}{n!}. \tag{5.9}$$

From (5.9), we can give Witt's type formula for multiple twisted Dirichlet's type $q$-Euler numbers.

Theorem 5.3.

Let $\chi$ be a Dirichlet's character with conductor $d(=\text{odd}) \in N$. For each $w \in T_p$, $h \in N$, and $n \in N \cup \{0\}$, we have

$$\underbrace{\int_X \cdots \int_X}_{h\text{-times}} \chi(x_1+\cdots+x_h) w^{x_1+\cdots+x_h}(x_1+\cdots+x_h)^n d\mu_{-q}(x_1)\cdots d\mu_{-q}(x_h) = E_{n,\chi,w}^{(h,q)}, \tag{5.10}$$

where $\chi(x_1+\cdots+x_h) = \chi(x_1)\cdots\chi(x_h)$ and

$$(x_1+\cdots+x_h)^n = \sum_{\substack{l_1+\cdots+l_h=n \\ l_1,\ldots,l_h \geq 0}} \frac{n!}{l_1!\cdots l_h!} x_1^{l_1}\cdots x_h^{l_h}. \tag{5.11}$$

From (5.10), we also obtain the sums of powers of consecutive multiple twisted Dirichlet's type $q$-Euler numbers as follows.

Theorem 5.4.

Let $X$ be a Dirichlet's character with conductor $d(=\text{odd}) \in N$. For each $w \in T_p$, $h \in N$, and $n \in N \cup \{0\}$, we have

$$E_{k,X,w}^{(h,q)} = \sum_{\substack{l_1+\cdots+l_h=k \\ l_1,\ldots,l_h \geq 0}} \frac{k!}{l_1!\cdots l_h!} E_{l_1,X,w}^{q} \cdots E_{l_h,X,w}^{q}. \tag{5.12}$$

Finally, we consider multiple twisted Dirichlet's type $q$-Euler polynomials defined by means of the generating functions as follows:

$$F_{w,X}^{q}(t,x) = \left(\frac{[2]_q \sum_{i=0}^{d-1}(-1)^{d-1-i} q^i X(i) w^i e^{ti}}{q^d w^d e^{td}+1}\right)^h e^{xt} = \sum_{n=0}^{\infty} E_{n,X,w}^{(h,q)}(x)\frac{t^n}{n!}, \tag{5.13}$$

where $|t+\log(qw)| < \pi/d$ and $\text{Re}(x) > 0$. From (5.13), we note that

$$\sum_{n=0}^{\infty} \underbrace{\int_X \cdots \int_X}_{h\text{-times}} X(x_1+\cdots+x_h) w^{x_1+\cdots+x_h}(x_1+\cdots+x_h+x)^n d\mu_{-q}(x_1)\cdots d\mu_{-q}(x_h)\frac{t^n}{n!}$$
$$= \sum_{n=0}^{\infty} E_{n,X,w}^{(h,q)}(x)\frac{t^n}{n!}. \tag{5.14}$$

Clearly, we obtain the following two theorems.

Theorem 5.5.

Let $X$ be a Dirichlet's character with conductor $d(=\text{odd}) \in N$. For each $w \in T_p$, $h \in N$, $n \in N \cup \{0\}$, and $\text{Re}(x) > 0$, we have

$$\underbrace{\int_X \cdots \int_X}_{h\text{-times}} \chi(x_1+\cdots+x_h) w^{x_1+\cdots+x_h}(x_1+\cdots+x_h+x)^n d\mu_{-q}(x_1)\cdots d\mu_{-q}(x_h) = E_{n,\chi,w}^{(h,q)}(x), \tag{5.15}$$

where

$$(x_1+\cdots+x_h+x)^n = \sum_{\substack{l_1+\cdots+l_h=n \\ l_1,\ldots,l_h\geq 0}} \frac{n!}{l_1!\cdots l_h!} x_1^{l_1}\cdots(x_h+x)^{l_h}. \tag{5.16}$$

Theorem 5.6.

Let $\chi$ be a Dirichlet's character with conductor $d(=\text{odd}) \in N$. For each $w \in T_p$, $h \in N$, $n \in N \cup \{0\}$, and $\text{Re}(x) > 0$, we have

$$E_{k,\chi,w}^{(h,q)}(x) = \sum_{\substack{l_1+\cdots+l_h=k \\ l_1,\ldots,l_h\geq 0}} \frac{k!}{l_1!\cdots l_h!} E_{l_1,\chi,w}^{q} \cdots E_{l_{h-1},\chi,w}^{q} \cdot E_{l_h,\chi,w}^{q}(x). \tag{5.17}$$

## REFERENCES

1. Kim T:Analytic continuation of multiple $q$-zeta functions and their values at negative integers. Russian Journal of Mathematical Physics 2004, 11(1):71-76.

2. Kim T:On the analogs of Euler numbers and polynomials associated with $p$-adic $q$-integral on $\mathbb{Z}_p$ at $q=-1$. Journal of Mathematical Analysis and Applications 2007, 331(2):779-792.

3. Kim T:A note on some formulae for the $q$-Euler numbers and polynomials. Proceedings of the Jangjeon Mathematical Society 2006, 9(2):227-232.

4. Kim T, Kim M-S, Jang L-C, Rim S-H:New $q$-Euler numbers and polynomials associated with $p$-adic $q$-integrals. Advanced Studies in Contemporary Mathematics 2007, 15(2):243-252.

5. Jang L-C:On a $q$-analogue of the $p$-adic generalized twisted $L$-functions and $p$-adic $q$-integrals. Journal of the Korean Mathematical Society 2007, 44(1):1-10.

6. Jang L-C, Kim S-D, Park D-W, Ro Y-S: A note on Euler number and polynomials. Journal of Inequalities and Applications 2006, 2006:-5.

7. Ozden H, Simsek Y:A new extension of $q$-Euler numbers and polynomials related to their interpolation functions. Applied mathematics Letters. In press

8. Ozden H, Simsek Y, Cangul IN:Remarks on sum of products of $(h,q)$v-twisted Euler polynomials and numbers. Journal of Inequalities and Applications 2008, 2008:-8.

9. Ozden H, Simsek Y, Cangul IN:Multivariate interpolation functions of higher order $q$-Euler numbers and their applications. Abstract and Applied Analysis. In press

10. Simsek Y:Twisted $(h,q)$-Bernoulli numbers and polynomials related to twisted $(h,q)$-zeta function and $L$-function. Journal of Mathematical Analysis and Applications 2006, 324(2):790-804.

11. Simsek Y:On $p$-adic twisted $q$- $L$-functions related to generalized twisted Bernoulli numbers. Russian Journal of Mathematical Physics 2006, 13(3):340-348.

12. Simsek Y:On twisted $q$-Hurwitz zeta function and $q$-two-variable $L$-function. Applied Mathematics and Computation 2007, 187(1):466-473.

13. Simsek Y:The behavior of the twisted $p$-adic $(h,q)$- $L$-functions at $s=0$. Journal of the Korean Mathematical Society 2007, 44(4):915-929.

14. Simsek Y, Kurt V, Kim D:New approach to the complete sum of products of the twisted $(h,q)$-Bernoulli numbers and polynomials. Journal of Nonlinear Mathematical Physics 2007, 14(1):44-56.

15. Carlitz L: $q$-Bernoulli numbers and polynomials. Duke Mathematical Journal 1948, 15(4):987-1000.

16. Cenkci M:The $p$-adic generalized twisted $(h,q)$-Euler- $L$-function and its applications. Advanced Studies in Contemporary Mathematics 2007, 15(1):37-47.

17. Cenkci M, Can M:Some results on $q$-analogue of the Lerch zeta function. Advanced Studies in Contemporary Mathematics 2006, 12(2):213-223.

18. Cenkci M, Can M, Kurt V: $p$-adic interpolation functions and Kummer-type congruences for $q$-twisted and $q$-generalized twisted Euler numbers. Advanced Studies in Contemporary Mathematics 2004, 9(2):203-216.

19. Cenkci M, Simsek Y, Kurt V:Further remarks on multiple $p$-adic $q$-$L$-function of two variables. Advanced Studies in Contemporary Mathematics 2007, 14(1):49-68.

20. Kim T: On Euler-Barnes multiple zeta functions. Russian Journal of Mathematical Physics 2003, 10(3):261-267.

21. Kim T:Power series and asymptotic series associated with the $q$-analog of the two-variable $p$-adic $L$-function. Russian Journal of Mathematical Physics 2005, 12(2):186-196.

22. Kim T:Multiple $p$-adic $L$-function. Russian Journal of Mathematical Physics 2006, 13(2):151-157.

23. Kim T:A note on $p$-adic $q$-integral on □$_p$associated with $q$-Euler numbers. Advanced Studies in Contemporary Mathematics 2007, 15(2):133-137.

24. Kim T: $q$-Euler numbers and polynomials associated with $p$-adic $q$ -integrals. Journal of Nonlinear Mathematical Physics 2007, 14(1):15-27.

25. Kim T:An invariant $p$-adic $q$-integral on □$_p$. Applied Mathematics Letters 2008, 21(2):105-108.

26. Kim T:On $p$-adic interpolating function for $q$-Euler numbers and its derivatives. Journal of Mathematical Analysis and Applications 2008, 339(1):598-608.

27. Kim T, Rim S-H:New Changhee $q$-Euler numbers and polynomials associated with $p$-adic $q$-integrals. Computers & Mathematics with Applications 2007, 54(4):484-489.

28. Kim T, Choi JY, Sug JY:Extended $q$-Euler numbers and polynomials associated with fermionic $p$-adic $q$-integral on □$_p$. Russian Journal of Mathematical Physics 2007, 14(2):160-163.

# INDEX

## F



## G



## H



## I



## J



## L



## M



## P



## Q



## R



## S

## T



## W



## Z